AF566884

The Tinbergen Legacy

The Tinbergen Legacy

Edited by

M.S. DAWKINS
Research Lecturer
Oxford University

T.R. HALLIDAY
Reader in Biology
Open University

and

R. DAWKINS
Reader
Oxford University

SPRINGER-SCIENCE+BUSINESS MEDIA, B.V.

First edition 1991

Originally published by Chapman & Hall in 1991
MyCopy version of the original edition 1991

Typeset in 10½ on 12pt Palatino by Mews Photosetting, Beckenham, Kent

DOI 10.1007/978-0-585-35156-8

A catalogue record for this book is available from the British Library

Library of Congress Cataloging-in-Publication data available

∞ Printed on permanent acid-free text paper, manufactured in accordance with the proposed ANSI/NISO Z 39.48–199X and ANSI Z 39.48–1984
www.springer.com/mycopy

Contents

Contributors

Gerard P. Baerends
Hoofdweg 265, 9765 CH Paterswolde, Netherlands

Nicholas B. Davies
Department of Zoology, University of Cambridge, UK

Richard Dawkins
Department of Zoology, University of Oxford, UK

Marian S. Dawkins
Department of Zoology, University of Oxford, UK

Juan D. Delius
Universität Konstanz, Konstanz, Germany

Timothy R. Halliday
The Open University, Milton Keynes, UK

Robert A. Hinde
Department of Zoology, University of Cambridge, UK

Felicity A. Huntingford
Department of Zoology, University of Glasgow, UK

John R. Krebs
Department of Zoology, University of Oxford, UK

Aubrey Manning
Department of Zoology, University of Edinburgh, UK

Michael H. Robinson
Smithsonian Institute, Washington DC, USA

Lary Shaffer
Department of Psychology, SUNY, Plattsburgh, USA

Introduction

RICHARD DAWKINS

A conference with the title 'The Tinbergen Legacy' was held in Oxford on 20th March, 1990. Over 120 of Niko Tinbergen's friends, family, colleagues, former students and people who had never met him in person converged at Oxford for what turned out to be a memorable day. To reflect the rather special atmosphere of the conference, we decided to begin this book with Richard Dawkins' opening remarks exactly as he gave them on that day.

Welcome to Oxford. For many of you it is welcome back to Oxford. Perhaps even, for some of you, it would be nice to think that it might feel like welcome home to Oxford. And it is a great pleasure to welcome so many friends from the Netherlands.

Last week, when everything had been settled except final, last minute arrangements, we heard that Lies Tinbergen had died. Obviously we would not have chosen such a time to have this meeting. I'm sure we'd all like to extend our deep sympathy to the family, many of whom, I'm happy to say, are at this meeting. We discussed what we should do and decided that, in the circumstances, there was nothing for it but to carry on. The members of the Tinbergen family that we were able to consult were fully in agreement. I think we all knew that Lies was an enormous support to Niko, but I think that very few of us really knew how much of a support she was to him, particularly during the dark times of depression.

I should say something about this memorial conference and what led up to it. People have their own ways of grieving. Lies' way was to take literally Niko's characteristically modest instruction that he wanted no funeral or memorial rites of any kind. There were those of us who were fully sympathetic to the desire for no religious observance, but who nevertheless felt the need for some kind of rite of passage for a man whom we had loved and respected for so many years. We suggested various kinds of secular observance. For instance, the fact that there was such musical talent in the Tinbergen family led some of us to suggest a memorial chamber concert with readings or eulogies in the intervals. Lies made it very clear, however, that she wanted nothing of the kind and that Niko would have felt the same.

So we did nothing for a while. Then, after some time had elapsed, we realised that a memorial conference would be sufficiently different from a funeral as not to count. Lies accepted this, and there came a time, during our planning of the conference, when she said that she hoped to attend the conference, although she later changed her mind about that, thinking, again with characteristic modesty and completely erroneously, that she would have been in the way.

It is an enormous pleasure to welcome so many old friends. It is a tribute to Niko, and the affection that his old pupils felt for him, that so many of you are here today, converging on Oxford from, in some cases, very far away. The list of people coming is a galaxy of old friends, some of whom may not have set eyes on one another for 30 years. Just reading the guest list was a moving experience for me.

We shall all of us have memories of Niko and of the group of his associates with whom we happen to be contemporary. My own begin when I was an undergraduate and he lectured to us, not at first on animal behaviour but on molluscs – for it was Alister Hardy's quaint idea that all the lecturers should participate in the 'Animal Kingdom' course which is one of the sacred cows of Oxford zoology. I didn't know, then, what a distinguished man Niko was. I think that if I had, I'd have been rather aghast at his being made to lecture on molluscs. It was bad enough that he gave up being a Professor in Leiden to become, by Oxford's snobbish custom, just plain 'Mr Tinbergen'. I don't remember much from those early mollusc lectures, but I do remember responding to his wonderful smile: friendly, kindly, avuncular as I thought then, although he must have been scarcely older than I am now.

I think I must have been imprinted on Niko and his intellectual system then, for I asked my college tutor if I could have tutorials with Niko. I don't know how he managed to swing it, because I don't think Niko gave undergraduate tutorials as a rule. I suspect that I may have been the last undergraduate to have had tutorials with him. Those tutorials had an enormous influence on me. Niko's style as a tutor was unique. Instead of giving a reading list with some sort of comprehensive coverage of a topic, he would give a single, highly detailed piece of work, such as a DPhil thesis. My first one, I remember, was a monograph by A.C. Perdeck, who I am happy to say is here today. I was asked simply to write an essay on anything that occurred to me as a result of reading the thesis or monograph. In a sense it was Niko's way of making the pupil feel like an equal – a colleague whose views on research were worth hearing, not just a student mugging up a topic. Nothing like this had ever happened to me before, and I revelled in it. I wrote huge essays that took so long to read out that, what with Niko's frequent interruptions, they were seldom finished by the end of the hour. He strode up and down the room while I read my essay, only occasionally coming to rest on whatever old packing case was serving him as a chair at the time, chain-rolling cigarettes and obviously giving me his whole attention in a way that, I'm sorry to say, I cannot claim to do for most of my pupils today.

As a result of these marvellous tutorials, I decided that I very much wanted to do a DPhil with Niko. And so I joined the 'Maestro's Mob', and it was an experience never to be forgotten. I remember with particular affection the Friday evening seminars. Apart from Niko himself, the dominant figure at that time was Mike Cullen. Niko obstinately refused to let sloppy language pass, and proceedings could be stalled for an indefinite period if the speaker was not able to define his terms with sufficient rigour. These were arguments in which everybody became engaged, eager to make a contribution. If, as a result, a seminar wasn't finished at the end of the two hours, it simply resumed the following week, no matter what might have been previously planned.

I suppose it may have been just the naïvcté of youth, but I used to look forward to those seminars with a sort of warm glow for the whole week. We felt ourselves members of a privileged élite, an Athens of ethology. Others, who belonged to different cohorts, different vintages, have talked in such similar terms that I believe that this feeling was a general aspect of what Niko did for his young associates.

In a way, what Niko stood for on those Friday evenings was a kind of ultra-rigorous, logical commonsense. Put like that, it may not sound like much; it may seem even obvious. But I have since learned that rigorous commonsense is by no means obvious to much of the world. Indeed commonsense sometimes requires ceaseless vigilance in its defence.

In the world of ethology at large, Niko stood for breadth of vision. He not only formulated the 'four questions' view of biology, he also assiduously championed anyone of the four that he felt was being neglected. Since he is now associated in peoples' minds with field studies of the functional significance of behaviour, it is worth recalling how much of his career was given over to, for instance, the study of motivation. And, for what it is worth, my own dominant recollection of his undergraduate lectures on animal behaviour was of his ruthlessly mechanistic attitude to animal behaviour and the machinery that underlay it. I was particularly taken with two phrases of his - ''behaviour machinery'', and ''equipment for survival''. When I came to write my own first book, I combined them into the brief phrase ''survival machine''.

In planning this conference, we obviously decided to concentrate on fields that Niko had been pre-eminent in, but we didn't want the talks to be only retrospective. Of course we wanted to spend some time looking back at Niko's achievements, but we also wanted people to pick up the torches that Niko had passed them, and run on with them towards the future.

Torch-running behaviour, in new and exciting directions, bulks so large in the ethograms of Niko's students and associates that planning the programme was a major headache. ''How on earth'', we asked ourselves, ''can we possibly leave out so-and-so? On the other hand, we have space for only six talks''. We could have limited ourselves to Niko's own pupils - his scientific children, but this would have been to devalue his enormous influence via grandpupils and others. We could have concentrated on people and major areas not covered in the *Festschrift* volume edited by Gerard Baerends, Colin Beer and Aubrey Manning, but that too would have been a pity. In the end, it seemed almost not to matter which half a dozen of Niko's intellectual descendants stood up to represent the rest of us. And perhaps that is the true measure of his greatness.

1 Early ethology: growing from Dutch roots

GERARD P. BAERENDS

In the Netherlands, between 1930 and 1940, ethology grew from what was originally seen as a pleasant and harmless hobby, to a new biological discipline, recognized by the academic world. Together with the Austro-German school, Dutch ethology came to play a leading role in this new study of animal behaviour. Its spectacular growth was due to the leadership of Niko Tinbergen, and I have been asked to give you here a kind of eyewitness report on how it all happened. Before doing this, I want to deal with some aspects of the cultural climate of the Netherlands in the first quarter of this century which were responsible for making that country one of the birthplaces of ethology and for facilitating Niko's development as one of its pioneers.

In the 1880s, coinciding with a growing awareness of the need for a more socially just society, cultural attitudes towards nature changed. Literature and the fine arts became increasingly interested in a realistic representation of nature. Writers and poets (Kloos, Albert Verwey, Gorter, Van Eeden), painters (Maris brothers, Israel, Mauve, Breitner, Wenckebach) and sculptors (Mendes da Costa) began to deal with landscapes, plants and animals in a style that took as much care with the correctness of naturalistic details as with the emotional impressions felt by the observer. Entirely new methods were developed for the teaching of children in primary schools, aimed at making them aware of the life and work of people in different communities and professions, and with particular emphasis on informing urban children about rural life.

Inspired by this atmosphere two schoolmasters, E. Heimans and Jac P.Thijsse, began in 1893 a lifelong cooperation aimed at enlightening the general public about the natural world around them. They began by writing a series of six popular books, each dealing with the life of plants and animals in a characteristic Dutch habitat and a field guide for identifying the more common animals and plants. In these books the authors guide the readers from one example to another, helping them to look for details in structure and behaviour and inspiring them to wonder about the mechanisms at work. The interest which was aroused by these books is reflected in the fact that by around 1915 they already had gone through three editions.

Furthermore, in 1896 Heimans and Thijsse founded a magazine named 'De Levende Natuur' (The Living Nature) in which naturalists were encouraged to describe their own observations. Between 1927 and 1986 Niko Tinbergen contributed nearly 100 papers to this magazine; moreover, from 1947 to 1978 he also acted as one of its editors. The greatest impact Heimans and Thijsse probably had was with the publication of a series of large illustrated albums of wildlife, which were published by the biscuit factory Verkade between 1906 and 1915. Several outstanding artists belonging to the naturalistic revival school I mentioned earlier contributed illustrations for these guides. Some of the pictures were packed with the biscuits and this practical exploitation of the 'collecting drive' that so many people have, proved to be a very successful way of conveying a message about natural history.

As a consequence of the increasing interest in natural history, naturalist societies were formed all over the country. These were mainly started by amateurs, but a few professional biologists joined them as well. A unique feature of the Netherlands - and one that in my opinion was very important for the development of ethology in our own country - was that young naturalists, from 11 to 23 years old, formed societies of their own, quite separate from those of adults. In 1920 most of these young naturalists' clubs came together into one national association devoted to the study of nature: the 'Nederlandse Jeugdbond voor Natuurstudie', abbreviated as NJN. The emergence of societies run entirely by young members themselves was a characteristic of this period and can be seen as one result of the movement for more social justice and democracy that took place at the end of the 19th century. It was part of a reaction against the heavy hand with which adult society always treated young people. This movement among young people originally arose in Germany and then rapidly spread to the Netherlands, but only there did it become instrumental in the promotion of interest in and knowledge about nature.

Niko Tinbergen was born in 1907 in The Hague, where he grew up in a family of one sister and four brothers. The cultural and intellectual changes taking place at the turn of the century had already made their mark on the life of the Tinbergen family. His father taught Dutch language and literature at a secondary school and had a great interest in the fine arts. All members of the family took a delight in close contact with nature, at that time still abundantly present in the dunes, woods and meadows around The Hague. The family used to spend their vacations in an area of heath and pinewood on an expanse of glacial sands near Hulshorst, about 150 km inland. Both parents were very hard-working and set high standards for their children, while at the same time giving them all possible freedom in their individual development. Niko was the first member of the family to take a deeper interest in wildlife. Later his youngest brother, Lukas, would follow his example. The other brothers turned to the physical sciences, while his sister became a linguist.

Niko's interest in nature mostly involved making his own observations and deductions rather than studying the scientific literature. He followed in the track of Heimans and Thijsse, whom he greatly admired. He particularly enjoyed the experience of secretly watching animals in the wild and capturing their behaviour in sketches and photographs. To a considerable extent these activities also satisfied his great passion for outdoor sports. Such pleasures were often shared with members of a small group of like-minded friends. When in 1920 the NJN was founded, Niko and his companions soon joined this association. While he was at secondary school, the NJN gave Niko the opportunity to develop his gift for passing his knowledge and experience on to others with clarity and incisiveness. A lecture he gave in 1929 to recruit new members for the association not only made me join the NJN, but also imprinted me on biology forever.

At school, Niko was unhappy with the institutionalized teaching programmes. He considered having to go to lessons a frustrating restriction on his freedom, but he wisely took care to make sure the marks on his school reports were always just above the level needed to avoid further curtailments on the time he could spend on nature-watching and sport. During this period two older people played an important role in encouraging Niko's wildlife observations and his thinking about the problems they raised. These were his biology teacher, Dr A. Schierbeek, and an outstanding amateur ornithologist, G.J. Tijmstra (maths teacher and headmaster of a drill school for obstinate boys!)

After finishing school in 1925, Niko was not at all certain that he should embark on an academic study of biology. He doubted whether the kind of biology that was at that time taught at the Dutch universities would help him in further developing his interest in nature in the wild. He seriously thought of becoming a sports teacher.

However, the problem of what to do when at 50 years of age he would have become too old for that job, stopped him. In order to open Niko's eyes to the potential of academic study, some of his older friends, who had recognized his gift for biological work, contrived a plan. They advised his parents to allow him to spend three months at the 'Vogelwarte Rossitten', on the Kuhrische Nehrung in the easternmost part of Germany, where high quality field work on bird migration was going on. The remedy worked; after his return from Rossitten, Niko enlisted for the biology course at the University of Leiden. There he found more understanding from his teachers than he had expected. Very importantly, he met Jan Verwey (Figure 1.1), a staff member 8 years his senior, who in 1924 made a now classic study of the

Figure 1.1 Dr Jan Verwey.

pair-formation behaviour of the grey heron. Jan Verwey was the son of the poet Albert Verwey, one of the pioneers of the literary renewal of the 1880s. He had spent his youth in the coastal dunes, where the family lived in a small village. Verwey fully shared Niko's devotion to wildlife studies, but in contrast to Niko he was well aware of the merits of the more conventional disciplines of biology and conscientiously kept track of what was appearing in the scientific literature. Verwey's example and support may well have been decisive in Niko's further development. They became, and always remained, great friends.

Niko was given a great deal of freedom to model his programme of study in any way he liked. He could, for instance, include his ecologically oriented field work on the food and feeding of raptors, in which his brother Lukas (8 years his junior) assisted him (Figures 1.2 and 1.3). In the summer of 1930 on the sands of Hulshorst, he began, entirely on his own initiative and with the aim of completing it for a PhD thesis, an experimental study of the way in which the digger wasp *Philanthus triangulum* manages to relocate its individual hole in the sand when it returns with prey for its larva. The earlier

Figure 1.2 Niko Tinbergen building a hide around 1925.

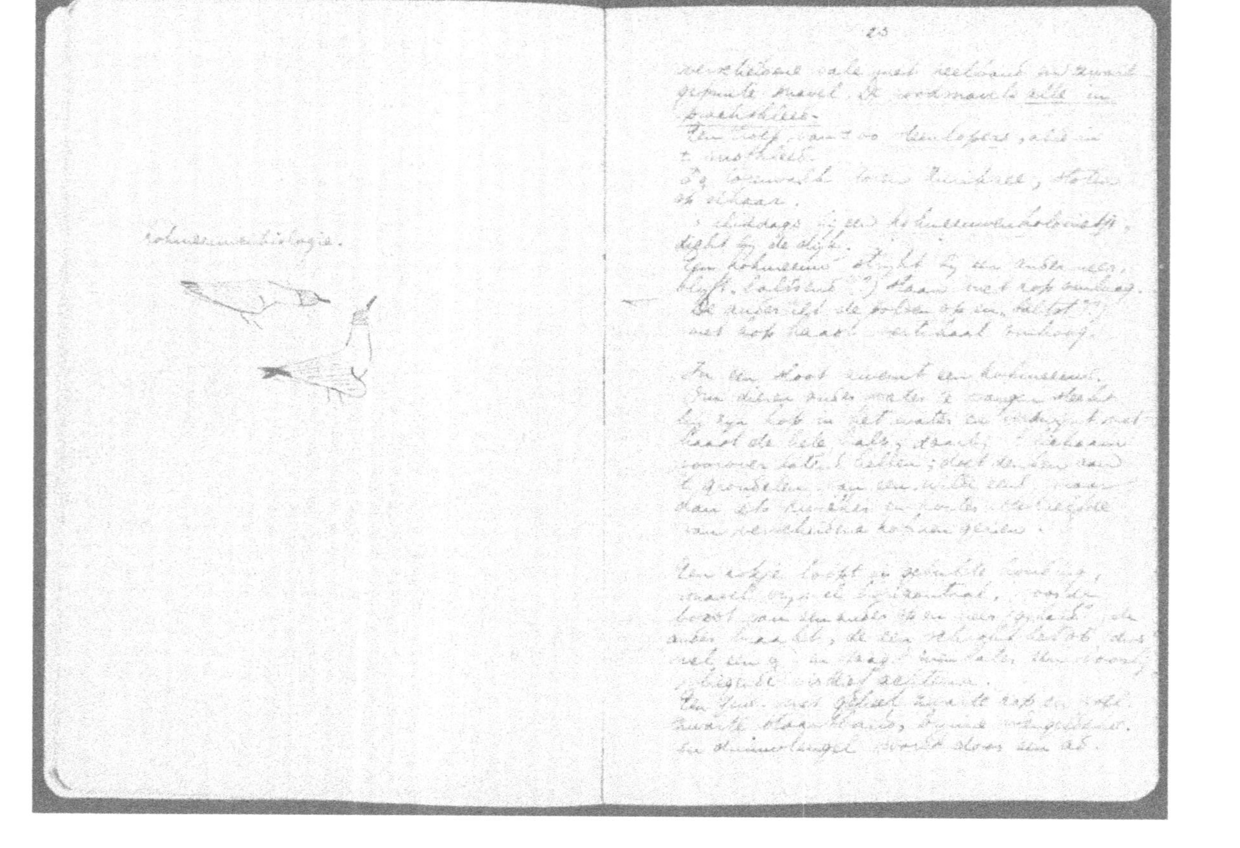

Figure 1.3a Extracts from Niko Tinbergen's notebook, which he started on 10 April 1930. These observations resulted in a paper in Ardea 21, 1932, pp. 1–13, written in Dutch, on comparative observations on some gulls and terns.

vechten v. visdiefjes.

duiken v. visdiefje.

[illegible]

Bij een punt aan de Scheldedijk [illegible] tientallen visdiefjes. Een vogel met [illegible] [illegible] op de dijk. Een ander [illegible] met ingetrokken, [illegible] hals, [illegible], [illegible].
Op [illegible] [illegible].
[illegible] [illegible] [illegible] houding; kop en hals schuin [illegible] hooggestrekt, [illegible].
[illegible] [illegible] [illegible] [illegible] [illegible] en staart [illegible].

Het duiken achter de punt [illegible] [illegible], [illegible] steeds schuin, [illegible] [illegible] [illegible] [illegible] [illegible], [illegible] [illegible] [illegible] [illegible] [illegible], [illegible] ook [illegible], [illegible] de vleugeltoppen [illegible].

♂♀ op [illegible] [illegible].
[illegible] de [illegible] [illegible] [illegible] rode.

[illegible] 's morgens [illegible], [illegible], [illegible] [illegible] [illegible], [illegible] [illegible] [illegible].
Een paartje [illegible] komt [illegible] [illegible] ♀ [illegible]. [illegible] ♂ [illegible], [illegible] op [illegible], het [illegible] steeds [illegible] [illegible] [illegible]. [illegible] [illegible] met uitgespreide vleugels [illegible]. [illegible] ♂ [illegible]?

studies made by Von Frisch and his co-workers on the orientation and homing of honey bees had inspired him to try this. However, before this study could be completed (data collection went rather slowly since the Dutch climate did not allow much more than 25 workable (i.e. sunny) days per summer) Niko was offered the opportunity to spend a year with a Dutch geophysical expedition to south-east Greenland. At that time this was such an exceptional opportunity that his official supervisor, Professor H. Boschma, allowed him to present the results he had obtained so far in an unusually slim thesis. He trusted Niko to return to this work and continue developing it later. In 1932, shortly before leaving for Greenland, Niko obtained his doctoral degree and

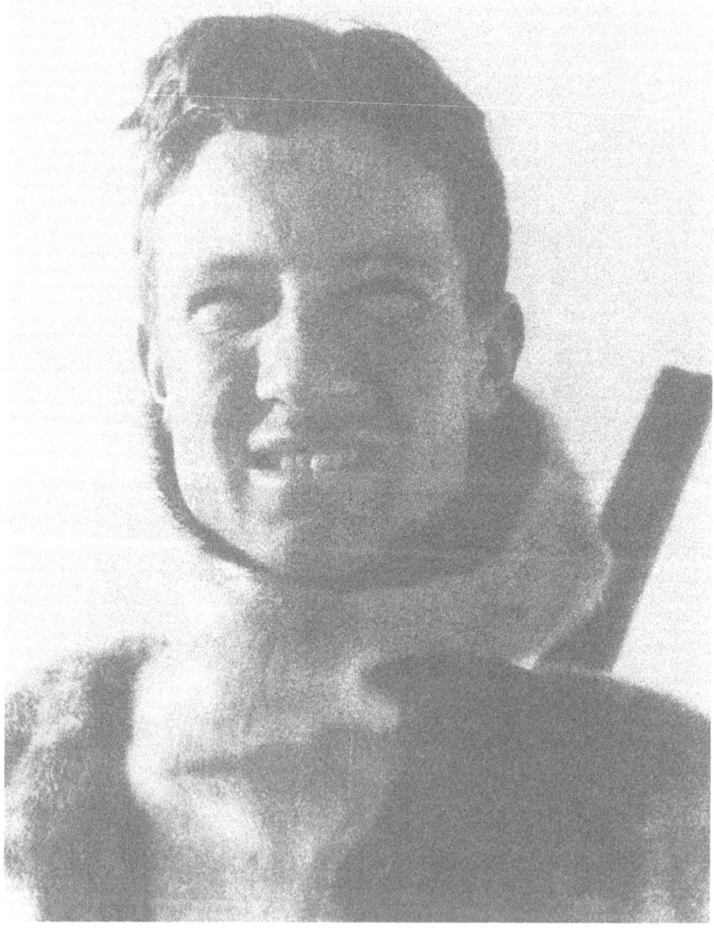

Figure 1.4 Niko Tinbergen in Greenland 1932–33.

married his girlfriend, Elisabeth Rutten, a sister of one of his birdwatching companions. Thus, the expedition was also to serve as Niko's honeymoon. Lies went with him and took part in all aspects of the field work. The Greenland year (Figure 1.4) offered Niko every opportunity to satisfy his desire to live amidst unspoiled nature. He made a thorough study of the reproductive behaviour of two bird species, the red phalarope and the snow bunting. He was also fascinated by the life of the Angmagssalikmut Eskimos and by the behavioural aspects of the symbiosis they had with their dogs.

After his return from Greenland in September 1933, Niko obtained an appointment as an assistant - the lowest-paid position in the academic hierarchy - at the Zoological Laboratory at Leiden. Officially charged with the task of running the elementary practical courses in invertebrate and comparative anatomy, Niko was also expected to develop ethology as a new part of the biology curriculum. This meant that he had the opportunity to prove that the hobby of nature-watching could be turned into a serious scientific endeavour. The director of the Zoological Laboratory, Professor C.J. van der Klaauw, himself a comparative anatomist with great interest in theoretical biology, gave him every possible support.

In Niko's view, the task of ethology was to study the phenomenon generally described as instinct - that is, the ability to perform species-characteristic behaviour adapted to survival in a specific ecological niche - with objective scientific methods. Following Julian Huxley, he emphasized that when watching behaviour, fundamentally different categories of questions could be distinguished, and that these should be kept carefully separated. In his early work Niko restricted himself to two of these questions, that of function (adaptation) and that of causation of the behaviour performed. Whereas for several behaviour patterns such as feeding activities or nestbuilding the function seemed to be evident, this certainly did not hold for the displays exhibited during social interactions. These displays had already attracted the attention of Niko and his friends, and were one of the features of behaviour that fascinated them most, as well as being wonderful subjects for photography. Consequently, the analysis of communication behaviour was an obvious choice for Niko's next research programme. The species most appropriate for study would be available in the neighbourhood of Leiden and easily observable in the wild or under semi-natural conditions. This led him to choose the herring gull for fieldwork on the nearby dunes and the three-spined stickleback - abundant in the ditches around the town and easy to breed in aquaria - for laboratory studies. In addition, during the summer months the work on the nest-orientation of *Philanthus* was resumed in the Hulshorst area.

Because of the prevailing economic crisis, financial resources were minimal. However, Niko did not see this as much of a handicap. In fact, apart from the Greenland expedition, he had always defrayed the expenses for his research from his own pocket. Fortunately, the first steps in ethology did not cost very much. Clear and disciplined thinking was the primary requirement, and at that time there was no real need for complicated and expensive equipment for data collection and processing (tools which in any case never interested Niko very much).

Figure 1.5 Jan Joost ter Pelkwijk.

When Niko started work in Leiden, only a few of the more advanced biology students were attracted to animal behaviour and fieldwork – in contrast to the younger students, many of whom had discovered biology through the NJN. Two students who arrived in the first year in September 1933 were Niko's younger brother Lukas and Jan Joost ter Pelkwijk (Figure 1.5). These two were of great help to Niko at the beginning because of their own enthusiasm and the stimulating effect they had on their colleagues. For several years both had shared many birdwatching activities with Niko and his other friends. Ter Pelkwijk was an extremely gifted man: he was highly original and unconventional; he was an excellent artist, wrote very well and had very broad interests. He loved being in the field, and had found it even more difficult than Niko to follow the institutionalized path of school work. Unfortunately, his contribution to ethology ended far too soon. During a stay in the Dutch East Indies, he was trapped by the war against the Japanese. In 1942 he was killed in action. In my opinion ter Pelkwijk's contribution to early stickleback studies was of vital importance, over and above his imagination and skill in devising dummies for experiments.

The Dutch university system provided excellent opportunities for developing new fields of experimental research. One opportunity came in the third year, when all biology students had to spend six weeks on experimental work. In 1936 ethology was officially accepted as a subject for such work and this probably led to the first ethology course in the world. At the beginning of the course the students were asked just to observe the behaviour of their animals and record it as carefully as possible, with pencil and paper. Following this introductory period of watching, they were encouraged to start asking questions, and thus to wonder about the behaviour they had observed. These questions were then critically discussed and where necessary corrected and refined. Finally, the students were invited to design and carry out experiments for testing their own hypotheses.

The feeling of working on their own project, which no-one had tackled before, was very stimulating to the students. Many of the projects were later to be used as stepping stones for approaching larger problems. The study of the role of the different features of male and female sticklebacks that released specific elements of their reproductive behaviour started in this way.

A second source of student help for Niko came with the larger research projects (lasting several months) that students were required to undertake before they could pass their final examinations. The problems suggested for these projects had often arisen from promising results obtained in the third-year course. Studies on the stimuli releasing and directing foodpecking in herring gull chicks, and begging in thrush nestlings, are examples of such projects. Undergraduate projects which had proved to be particularly promising then provided a good basis for longer PhD research.

Niko expanded the manpower of his ethology group by inviting undergraduates to help as unpaid assistants with the field studies. It was a great privilege to serve as such a 'slave'. We gladly devoted the greater part of our vacations, and in spring also gave up many hours between sunrise and the start of our lectures (which were then in constant danger of being missed). For me, it meant the start of my work on stimulus selection in the herring gull as well as on the reproductive behaviour of the digger wasp *Ammophila*. Niko took part in the fieldwork as much as his duties at the laboratory allowed, setting an example for working efficiently and in a well disciplined way, not allowing the hardships and blessings of the prevailing field conditions to get in the way. Unobtrusively, he taught us to keep our eyes open all the time to phenomena not directly related to our own project. Moreover, working and living in the field together provided excellent opportunities for personal contact between teacher and students and for informal education. This was particularly true when – as for the insect studies at Hulshorst – there were long periods of camping out in the field. Here, students also learned how to behave in the field, with respect for the flora and fauna as well for the people, particularly the warden and owners living and working in the study area.

Niko always tried to base his research on knowledge of the environment, behaviour and morphology of the animals he was studying as well as those of related species. Because of this attitude, he had an aversion to most of the work of the behaviouristic schools. In the early 1930s his teaching and planning with respect to the functional aspects of behaviour was often inspired by the writings of ornithologists such as Huxley, Selous and Howard, or insect-watchers like Fabre and Ferton. In his own country, the impressive knowledge and experience of Frits Portielje, a self-taught animal psychologist at the Amsterdam Zoo, was a source of inspiration to him. Nevertheless, Niko strongly rejected Portielje's view – fervently defended by the vitalistic animal psychologist Bierens de Haan – that the phenomenon of instinct would not be accessible to analysis. For the study of causal mechanisms, the work of behavioural physiologists such as Von Uexküll and Kühn, but above all that of Karl Von Frisch, was even more important. Von Frisch can be said to have founded the art of making an animal answer questions; this was exactly what Niko needed and wanted to extend his own experiments. Von Frisch's approach inspired various studies of the Leiden group on how animals perceive and recognize those stimulus situations in their environment that are of biological significance to them. Interest in applying ethological findings to human behaviour did not exist at that time. On the contrary, the emphasis was rather on the apparent differences between animal and human behaviour.

It was not until the mid-1930s that Konrad Lorenz' ideas about the nature of instinctive behaviour came to the Leiden group and began to influence the way they thought as well as the structure of their research programme. The theoretical concepts of 'fixed action patterns' (FAP) and 'innate releasing mechanism' (IRM) postulated by Lorenz provided a challenge as to how they were to be experimentally verified. Experimental testing of the IRM concept was entirely in line with studies already going on in Niko's group. The FAP concept, however, inspired him to extend his interest to a new area, the problem of the evolution of behaviour, and thus to ask a third fundamental biological question.

Lorenz and Tinbergen first met in person in November 1936, when Lorenz visited Leiden on the occasion of a symposium dedicated to the concept of instinct organized by Professor Van der Klaauw. In the spring of 1937, the Tinbergens spent several months with Lorenz in his home research station at Altenberg, near Vienna. During this stay Lorenz learned to appreciate Niko's gift for experimentation as together they studied the nature of the egg-retrieving activity of the greylag goose and the stimulus situation evoking the flight responses of geese and fowl chicks to aerial predators. Niko and Konrad differed greatly in personality but their attitude towards nature was the same. They shared a predilection for living with their animals – Niko preferably as a non-participating hidden observer and Konrad as an adopted alien member and protector – and they both obtained great satisfaction from the sense of understanding them. They also had a similar sense of humour. Both were impressive personalities, but of very different kinds. Konrad was a great talker, knowledgeable in a wide field and always bursting with ideas, which however he was not particularly keen to verify systematically. Niko was a good listener, who always tried to put what he heard or saw into a clearly formulated framework, accessible to critical verification and therefore open to improvement in the future. Consequently their contributions to ethology were complementary. They mutually appreciated this and recognized that they needed one another. It made them become and remain close friends.

Lorenz' concept of the FAP influenced Niko's thinking and his research programme in two ways. First, the characteristics Lorenz had attributed to FAPs demanded experimental verification. In particular the assertions that each FAP possessed its own internal autonomous impulse-generating mechanism and that the occurrence of an FAP was not subject to superimposed control mechanisms common to different FAPs, needed to be tested. Second, the notion that the concept of homology could be applied to behaviour as well as morphology opened the door to a study of the factors underlying the evolutionary radiation of behaviour, and the ways in which a behaviour repertoire can in the course of phylogeny, be extended by the addition of new elements.

As to the nature of FAPs, the experiments with sticklebacks by Niko and his co-workers forced them to reject Lorenz' view that mechanisms for control of different FAPs did not overlap. They found that specific stimulus situations could exert a common influence on the threshold for the release of different FAPs. Since the threshold change often persisted after this triggering stimulation had gone, it was concluded that internal factors inducing a particular behaviour state are involved in this common control of several FAPs. Behavioural states of different levels of integration could be distinguished and evidence was obtained that between states of equal level, inhibitory relationships existed. The conclusion was drawn that a hierarchical structure is basic to the way behaviour is organized. The various behavioural states – for which Niko claimed the name 'instincts' – tended to subserve different survival functions. In the hands of the Leiden ethologists (in particular Van Iersel and Sevenster) the exploration of the structure of the behavioural organization by means of 'black-box' analysis became increasingly sophisticated.

If the concept of homology is applied to behaviour, comparative studies of FAPs in different species can be expected to provide answers to questions about phylogeny – that is, from what antecedents often bizarre-looking displays may have been derived in the course of evolution. Niko observed that often a display, either parts or all of it, could be seen as homologous to other activities serving a direct, instrumental, function in a different context. The way in which such activities could have become involved in communicative displays seemed to be revealed when in a number of cases displays were found to be composed of incomplete elements of attack and escape behaviour, blended into a compromise. Such displays could be thought of as resulting from simultaneous activation of tendencies to attack and to escape, likely to occur when conspecifics, unknown to each other, meet. At that time, in the abscence of the game theory approach, it seemed obvious that it would be adaptive for animals to evolve displays by which they could appease their opponents. Studies of considerable sophistication supported the view that displays (which often resembled so called 'displacement activities') could have evolved as a result of the interaction between incompatible behaviour states. Thus, in this 'conflict hypothesis' specific properties of the structure of the behavioural organization were thought to provide a basis for the evolution of new activities. The hypothesis can also help to understand the variation and adaptive radiation of displays.

This conceptual framework had become one of the guiding principles of the research programme of the Leiden ethologists when the war broke out. Most of the research and the official teaching had to be stopped, but at least initially, work on data that had already been collected and on committing results and ideas to paper still went on. However, contact between the members of the group became gradually more difficult, especially when, in September 1942, Niko was taken hostage by the occupying Germans. He was put in a camp where prisoners were in constant danger of being shot in reprisal for attacks by the Dutch Underground forces (Figure 1.6).

After the liberation in 1945, the University of Leiden was reopened and the work of the ethological department resumed on a prewar basis. Niko, who between 1933 and 1939 had been gradually promoted to higher academic ranks, was now nominated for a full professorship. This promotion, however, meant that he would be charged with the directorship of the Zoological Laboratory. He had good reasons for fearing that the administrative duties involved would seriously handicap him in his principle aim, the promotion and further development of ethology. Lecture visits to America and Britain just before and just after the war had made Niko aware of the rising interest in ethology in these countries. He had concluded that for the discipline to become really established in the English-speaking world, the emigration of continental ethologists would be helpful or even essential. Convinced that the survival of ethology in the Netherlands was secured, in 1949 Niko accepted an invitation to move to the University of Oxford, to which he felt especially attracted by the high level of research into evolutionary questions.

In Oxford, Niko initially continued his previous research programme and many PhD students were attracted to ethology. Some of them embarked upon studies inspired by the concept of the IRM but under the influence of Lehrman's sharp critique of the Lorenzian 'innate vs learned' dichotomy were careful to give due attention to the role of experience in the development of an animal's knowledge of its environment. The need for a fourth 'why' question in ethology, the ontogenetic one, emerged clearly, but the major part of the research programme in the first decade of Niko's Oxford period still dealt with

Figure 1.6 A drawing by Niko Tinbergen of a seminar held in the hostage camp in 1944. Niko felt that he learned a great deal about drawing portraits from lessons given by one of the other hostages, the artist Karel Van Veen.

studies on the evolution of behaviour, with comparative studies on the displays of different gull species forming the core.

The possibilities of giving causal and phylogenetic interpretations of displays on the basis of ethological hypotheses about the organization of behaviour were so promising and fascinating that it tended to be forgotten that these hypotheses still needed to be critically verified and more explicitly worked out. Therefore, quantitative methods applying the ethological black-box analysis had to be developed and techniques for recording and processing data had to be improved. Care had to be taken to avoid premature conclusions about the physiological basis of the causal processes involved, an error early ethologists had sometimes committed in order to make their discipline look more respectable. The type of work required here did not appeal to Niko. Moreover, in the meantime his studies on the adaptive radiation of displays had made him aware of another urgent need, that for empirical assessment of selection pressures showing natural selection at work. Filling this gap was much more to his liking; here measurements had to be obtained in the field, preceded by watching and wondering. Consequently, after 1960, functional research on the adaptive nature of behaviour came to dominate Niko's research programme. This proved to be a step towards the integration of ethology and ecology, from which in the 1970s the discipline of Behavioural Ecology emerged. It also led to his involvement in the wildlife research of the Serengeti Research Institute in Tanzania, which gave Niko a second and last great opportunity to feel himself part of unspoiled nature.

2

Studying behavioural adaptations

NICHOLAS B. DAVIES

When I was a schoolboy, every Sunday I used to go to the local pinewoods and sand dunes to spend the day watching birds. I wrote an account of the species I saw for the school natural history competition and won a book by Niko Tinbergen (1953), 'The Herring Gull's World'. (Some time later I discovered that my prize was automatic as mine was the only entry.) Reading this book was a revelation. It revealed to me for the first time a whole new way of asking questions about natural history, a subject which until then I had assumed to consist simply of making species lists. Later, during my undergraduate days, the impression I got was the rather depressing one that you had to be incredibly clever to do research and that new ideas would emerge only from long hours in the laboratory or library. It was refreshing to return to Tinbergen's book, with its emphasis on patient field observation, which gave the encouraging idea that any birdwatcher could make a great discovery if only he had a spare afternoon and a pair of binoculars.

In this essay I shall consider how Tinbergen's legacy has influenced the way we now study behavioural adaptations, particularly the approach adopted in modern day 'behavioural ecology'. Is what we do now so very different from the work of Tinbergen and his students in the 1950s and 1960s? To what extent are Tinbergen's ideas relevant to a student embarking on behavioural research today?

LESSONS FROM TINBERGEN

I think that Tinbergen has had two important influences.

Asking questions

A mark of the most creative scientists is not only the kind of answer they provide to problems but the new kinds of questions they ask. As Tinbergen wrote in his introdution to 'The Herring Gull's World': 'Awareness of ignorance is in itself the result of some sort of understanding, the understanding and knowledge of problems to be solved'. The questions Tinbergen asked were interesting because they stemmed from a good knowledge of the animal and its world. For example in 1963 he wrote: 'It took me ten years of observation to realize that the removal of the empty eggshell after hatching, which I had known all along the black headed gulls to do, might have a definite function'. Then followed the famous experiments with broken shells laid out in artificial nests, which showed that their white interior attracted predators to an otherwise camouflaged nest, demonstrating that eggshell removal was indeed adaptive (Tinbergen *et al.*, 1963).

Tinbergen also had a knack of asking questions which could be answered; he framed questions clearly so that the reader would think 'I know how to answer that'. As Lorenz commented in his foreword to 'The Herring Gull's World' 'Tinbergen knows exactly how to ask questions of nature in such a way that she is bound to give clear answers'.

Tinbergen's (1963) distinction between the four kinds of questions (causation, function, development and evolution) is still very relevant today and his 1963 *Zeitschrift* paper still stands as a valuable reference on student reading lists.

Answering questions

Tinbergen not only had a genius for asking good questions, in contrast to Lorenz he also followed up his intuition by rigorous testing of his hypotheses. Here his main contribution was to show that the field can be used as a laboratory for observation and experiments. Tinbergen was not the first to do field experiments of course, but he was one of the first to collect quantitative data, to do careful controls, and to influence others by his example that field experiments were valuable for dissecting both cause and effects of behaviour patterns. By controlling variation ourselves in an experiment we can eliminate the possibility that another variable correlated with the feature understudy is the cause or effect of an event, e.g. is it the red spot on the parent's bill that causes pecking by the chicks, or some other feature? Experiments can be used to increase the range of natural variation to create circumstances that rarely or never occur. Andersson (1982) used this to good effect when considering the question of why male widowbirds *Euplectes progne* have such extraordinarily long tails. He showed by experimentally elongating tails not only that males with longer tales attracted more females, but also that increasing tail length beyond that normally observed increased the number of mates. Thus female preference selects for still longer male tails and the observed tail length must represent a balance between the opposing forces of natural selection and sexual selection.

Tinbergen's emphasis on the importance of the 'animal's world' also showed the value of the comparative method. Different species are expected to have adaptations relevant to their different worlds, as shown for example by the differences in parent and chick behaviour between the black headed gull *Larus ridibundus* and the kittiwake *Rissa tridactyla*, which reflect differences appropriate to their different nest sites, on the ground and on cliffs respectively (Cullen, 1957). The comparative method is a powerful tool used today to study adaptation (Clutton-Brock and Harvey, 1984), in effect using the way selection has designed species as the results of 'natural experiments' where evolution has had to solve problems posed by differences in ecology.

All these ideas, derived from Tinbergen, form an important basis for current research. What, then, is different? Differences become obvious if one compares any modern textbook on animal behaviour with Tinbergen's 'Social Behaviour in Animals' (2nd edn 1964, reprinted 1990). I shall highlight four differences in the emphasis of research today.

DIFFERENCES IN CURRENT RESEARCH

Measuring the success of behavioural design

When Tinbergen referred to the 'survival value' of behaviour, he clearly recognized that this meant 'fitness consequences'. He asked how the behaviour helped the animal in maintaining itself and its offspring. There have been two major changes in current thinking about design success.

First, thanks to Hamilton (1964), we now realize that behaviour can be favoured by selection not only because of its beneficial effects on descendent kin (offspring) but also because of beneficial effects on non-descent kin (e.g. siblings). Second, the development of the theory of evolutionarily stable strategies (Maynard Smith, 1982) has shown that there may not be a single 'best' design for a behaviour pattern.

Where the success of a strategy is frequency-dependent, the stable outcome of selection may be for variability in the population (e.g. variability in male dungfly *Scatophaga stercoraria* waiting times at a cowpat; Parker, 1970).

Studying trade-offs

Tinbergen emphasized that there are conflicts between different selection pressures and so characters may represent a compromise. For example, black-headed gulls do not remove eggshells at once but wait until one or two hours after hatching. Two factors may favour a delay. First, the chick needs time to free itself completely from the shell: too early removal may hurt the chick. Second, newly hatched chicks are wet and easily swallowed by predatory neighbours; by waiting until the chick has dried out and become fluffy, and so less easily swallowed, the parent decreases the chance of that chick being predated during the brief absence to remove the shell. The result is a compromise: remove the shell 'not too early but not too late' (Tinbergen *et al.*, 1963).

This idea is very influential today, with mathematical models used to measure trade-offs quantitatively and so predict optimal design (Stephens and Krebs, 1986). Design features of behaviour are often linked to life history theory to predict, for example, how much a parent should risk for its young at the expense of its own survival (Regelmann and Curio, 1986; Dijkstra *et al.*, 1990).

Individual differences

One of Tinbergen's main interests was to understand why different species behaved in different ways. He realised that to understand these differences you had to look at how behaviour was of advantage to the individuals of the species, so he studied individuals to reveal the significance of species-specific behaviour patterns, for example removal of eggshells, fear of cliffs, and crypsis (Tinbergen, 1974). David Lack too was especially interested in species characteristics. His book on the robin *Erithacus rubecula* (Lack, 1965) is all about what the species does: why it defends territories, what its life-span is, and so on. Likewise, Lack's interest in clutch size was to understand why selection has favoured a particular average clutch size for the great tit *Parus major* and the swift *Apus apus* (Lack, 1966). His interest in breeding systems was to understand questions such as why some species are monogamous while others are polygamous (Lack, 1968).

While these remain important problems for study, much current work is aimed at understanding individual differences within a population. Long-term studies are made, often over individuals' lifetimes, to elucidate why some individuals defend territories while others do not, why some individuals lay larger clutches or have more mates than others. Individual differences are the raw material for natural selection and studying them can be useful both for understanding adaptation and selection in progress (Grafen 1988).

A good example of how a long-term study of individual differences is sometimes needed to understand the adaptive significance of design features is the study by Clutton-Brock *et al.* (1982; 1984) of maternal behaviour in red deer *Cervus elaphus*. Individual differences in birth weight and early growth affect adult size and reproductive success, especially in males. This probably explains why hinds allow sons to suckle almost twice as frequently as daughters, even though this is costly to the mother. Rearing a son increases the chance that the mother will die the following winter and reduces her chance of breeding successfully the next year even if she does survive. The lifetime reproductive success of a son increases more rapidly with the mother's quality (related to dominance rank, which is correlated with body size) than does the lifetime reproductive success of a daughter. All females breed, even poor-quality individuals. However, because of intense competition for harems, only good-quality males can gain mates. The most successful males, which gain large harems, have the highest lifetime reproductive success of all individuals. Thus dominant hinds, able to produce good-quality sons,would maximize their reproductive success by biasing their sex ratio towards male offspring. Subordinate mothers, unable to raise good male competitors, would do better by producing daughters. These predicted biases in the sex ratio were, in fact, observed.

None of these results could have been established without following individuals over their lifetime. The most important contribution of longitudinal studies (e.g. Clutton-Brock, 1988; Stacey and Koenig, 1990) is that they allow us to relate events at one stage of an animal's life history to its survival and reproductive success at subsequent stages. The costs and benefits of an individual's actions are often delayed and there is no way of measuring these realistically unless individuals are followed through time.

Social behaviour: conflicts of interest

In his 1964 book on social behaviour, Tinbergen's main interest was in how cooperation was achieved between individuals through signalling. Thus he was mainly concerned with causal explanations of social interactions, for example how a stickleback's or gull's displays caused others to approach or retreat, how signals from the parent caused offspring to beg and how signals from the offspring caused parents to feed them. By contrast, the emphasis in current research is more on the conflicts of interest between individuals in relation to how they might best maximize their reproductive success. Recognition of conflicts of interest has played a seminal role in modern interpretations of mate choice, mating systems and parent–offspring interactions (Trivers, 1972; 1974).

Some of Tinbergen's interpretation of behaviour in social groups was group selectionist. For example, he interpreted the mobbing of a sparrowhawk by a group of wagtails as behaviour which, although of danger to the individual, was advantageous to the group. He argued that 'only groups of capable individuals survive – those composed of defective individuals do not, and hence cannot reproduce properly. In this way the result of cooperation of individuals is continually tested and checked, and thus the group determines ultimately, through its efficiency, the properties of the individual'. (Tinbergen, 1964; new edition 1990). In his later writings, Tinbergen (1973) argues clearly against the idea of group selection, emphasizing that cooperation comes about because individual participants gain an advantage. However he did not study the conflicts between individuals which underlie even apparently cooperative enterprises such as group mobbing or breeding.

One of the reasons for these differences in emphasis is that behavioural ecologists have focused mainly on the fitness consequences of behaviour, whereas Tinbergen gave equal time to studies of both causation and function. He was interested not only in why the black-headed gull removed eggshells, in the sense of what good it did them but also in the stimuli which elicited removal (Tinbergen *et al.*, 1962). For example, he showed that various characteristics such as a 'thin edge' elicited removal (a whole egg with a flange glued to it was removed), whereas others such as 'shape' did not (a half egg filled with plaster of Paris was rolled back into the nest).

Behavioural ecologists, by contrast, have largely ignored mechanism. In the rest of this essay I shall discuss one of the 'hot topics' in behavioural ecology, namely helping at the nest, to show the relevance of three aspects of the Tinbergen legacy to today's research: a) the importance of distinguishing different kinds of question about behaviour; b) the usefulness of studying both mechanism and function together, and c) the role of field observation and experiments.

LINKING CAUSAL AND FUNCTIONAL STUDIES OF HELPING AT THE NEST

In many species of birds the young are fed not only by their parents but also by one or more 'helpers'. The question arises, what benefit does a helper gain from such altruism? Why don't helpers go off and rear their own young instead of helping others to breed? Field studies have revealed that in many cases the helpers are previous offspring of the breeders. They remain at home often because the habitat is full, so there are no breeding vacancies. Because the young in the nest are sibs of the helper, the helper can increase its genetic fitness by feeding them (Brown, 1987; Stacey and Koenig, 1990). The traditional behavioural ecology view of helping is therefore that 'genes for helping' have spread by kin selection.

This interpretation has recently been criticized by Jamieson and Craig (1987; see also Jamieson, 1989). They propose that individuals have simple 'provisioning rules' favoured in the context of parental care. Such a rule might be 'Feed begging chicks in my territory'. When the habitat is full and juveniles remain in their natal territory, they encounter nestlings when their parents have another breeding attempt and the presence of the begging chicks elicits provisioning.

The standard reply to Jamieson and Craig has been that theirs is a causal explanation of helping (i.e. what elicits feeding) whereas behavioural ecologists have been interested in a different level of analysis, namely its functional significance (Sherman, 1989; Ligon and Stacey, 1989, Koenig and Mumme, 1990). However, I do not think that this is the key issue. Jamieson and Craig are suggesting that 'helping' is not a trait. Rather 'provisioning' is the trait and helping is simply a by-product of a rule favoured in the context of parental care. Their point is that helping behaviour may not have arisen because of the spread of a 'gene for helping'. Just because helping brings a benefit does not mean that selection has specifically favoured it as a trait. In his 1963 paper, Tinbergen highlights this problem when he refers to a difficulty 'caused by our habit to coin terms for major functional units and treat them as units of mechanism'.

To examine Jamieson and Craig's hypothesis we need to look at the design of helping behaviour, to understand alternative mechanisms, before we can ask sensible functional questions. Our functional questions need to be of the form 'Why does this animal do *x* when it could do *y*?' Selection chooses between alternative mechanisms; which alternatives are available is important for any functional argument. So the key question is: is 'not helping' an alternative? Or do individuals blindly follow a crude provisioning rule? Understanding the mechanism will tell us whether we should be measuring the costs and benefits of 'provisioning' or of 'helping'.

Some recent studies have shown that individuals do not always provision chicks but rather do so only when they are likely to enjoy a genetic gain from doing so. For example, white fronted bee-eater *Merops bullockoides* helpers almost always only help feed the offspring of close relatives and furthermore, given the choice between helping close versus more distant kin they almost always choose to help the closer kin. Individuals who do not have close relatives in the colony tend not to help (Emlen and Wrege, 1988). Such intricacy of behavioural design shows that the question 'why help?' is a sensible one to ask, one that begs a functional interpretation.

Two other studies also argue against Jamieson and Craig's view and provide an interesting comparison.

Acorn woodpeckers (*Melanerpes formicivorous*)

This species is a communal breeder, with several related females laying in the same nest and several related males breeding in the group (Koenig and Mumme, 1987). One male is dominant and may gain most of the mating access to the females. Limited data show there can be multiple paternity in a brood. Often all the breeding males help to feed nestlings.

A male may thus gain two kinds of genetic benefit from feeding nestlings. First he may gain 'direct fitness' benefits (Brown, 1987) from feeding his own offspring in the brood. Second, he may gain 'indirect fitness' benefits from feeding non-descendent but related offspring, namely those sired by other breeding males (to whom he is related - co-breeding males may be father and son, or brothers for example). Koenig (1990) asked 'what causes a male to feed the chicks?' He performed a neat field experiment which involved asking the woodpeckers a clear question. He removed a male during egg laying, so that he did not have the chance to father any of the chicks. He then replaced the male during incubation to see whether he fed the young. The experiment thus allowed Koenig to test whether the 'indirect fitness' benefits of helping were sufficient to cause feeding.

The data showed that when dominant males were removed, they destroyed the clutch on their return to the territory, thus forcing a re-nest. In control experiments, where dominant males were removed during incubation after all the mating was over, they never destroyed the clutch; this shows that it was their absence during the egg-laying period that induced clutch destruction, not the removal *per se*. By contrast, subordinate males did not destroy the clutch, even though they had the opportunity to do so. Some subordinates, at least, then went on to help feed the nestlings (Koenig, 1990). These results suggest that for dominants the fitness gain from forcing a re-nest may be greater than the indirect fitness gain from feeding non-descendant kin, while for subordinates the reverse may be true. It seems likely that dominants are usually able to gain greater paternity and so the payoff to them of a re-nest may be more profitable. The next step must be to measure paternity bias and to calculate these payoffs to see if these different decision rules for helping in dominants and subordinates indeed maximize fitness.

The main point from this example is that by revealing differences in what causes feeding of nestlings by dominants and subordinates, the study can now go on to ask interesting functional questions.

Dunnocks (*Prunella modularis*)

Dunnocks are often polyandrous, with two males sharing a female. Unlike the woodpeckers above however, the males are not close relatives (Davies, 1990). DNA fingerprinting shows that broods are sometimes sired entirely by one of the males (usually the dominant, or alpha, male) while sometimes paternity is shared between alpha and beta male. The interesting result is that beta males are more likely to help feed the chicks if they have some paternity (Burke *et al.* 1989). Do the males have an equivalent of DNA fingerprinting to guide their behaviour, or do they rely on simple rules? The data support the latter view. First, beta males sometimes help feed broods when they have no paternity. Sometimes alpha and beta males cooperate to feed a single chick (clearly they cannot both be the father). Second, when the chicks fledge they are usually divided among the parents, with each male taking sole care of some of them until they reach independence. There is no tendency for a male to pick out his own offspring for care. How then does the relationship between paternity and chick feeding come about? Behavioural observations show that a male's mating access to the female predicts paternity reasonably well and that males use their degree of mating access to determine whether they feed the young (Burke *et al.*, 1989). It is a simple rule which works quite well in the sense that it results in a male provisioning broods where he has some paternity.

This example provides a nice contrast to the acorn woodpecker, where a subordinate male will feed chicks even if he has no mating access to the female. In the woodpeckers the males are close relatives. In dunnocks they are not, so there is no kin-selected benefit for a beta male dunnock to help feed an alpha male's offspring. Thus the different mechanisms leading to chick feeding in the two species each make good functional sense.

These studies of helping at the nest illustrate the importance of linking studies of mechanism and function. Behavioural ecologists interested in function should ask questions of the form 'why have particular mechanisms been favoured in some species while different mechanisms have been favoured in others?' The subject is in part a comparative study of mechanism in relation to ecological circumstances.

IS BEHAVIOURAL ECOLOGY DIFFERENT FROM ETHOLOGY?

In his 1963 paper, Tinbergen began with the following warning: 'I believe that if we do not continue to give thought to the problems of our overall aims, our field will be in danger of either splitting up into seemingly unrelated subsciences, or of becoming an isolated "ism".' This theme was taken up by Wilson (1975) who predicted that ethology would be 'cannibalized by neurophysiology and sensory physiology from one end and sociobiology and behavioural ecology from the other'.

In part, Wilson's prophesy seems to have been fulfilled. We now have separate societies and journals in neuroethology and behavioural ecology, which appears to leave ethology dwindling in between. However I do not think this gives a fair picture of current trends. Although in its early days behavioural ecology tended to ignore mechanism, it is interesting to note a rising interest in questions of causation and development. For example, optimal foraging studies began by trying to understand why animals preferred particular patches or prey. Recent studies have turned to questions of how foragers assess patch quality, how they assess travel costs and other problems concerning mechanism (Stephens and Krebs, 1986). What animals can do will be constrained by the mechanisms available to them and obviously we need to understand these mechanisms. Students of cooperative breeding and mating systems are likewise becoming more interested in studies of causation and development. What causes individuals to feed nestlings? How do helpers recognize kin?

In my study with Michael Brooke of cuckoos *Cuculus canorus* and their hosts we were mainly interested in how the cuckoo tricked the host and how hosts had evolved counteradaptations. One of the discoveries, using Tinbergen-type experiments with model eggs, was that many hosts rejected eggs which were different from their own colour and pattern, thus explaining why the cuckoo has evolved a mimetic egg for these hosts (Davies and Brooke, 1988; 1989a and b). This raises an interesting developmental question, namely how do hosts know what their own eggs look like? There is an interesting developmental question to ask about the cuckoos: how does the cuckoo come to select the right host, namely the one for whom its egg is a good match? Some hosts, like the dunnock, show no discrimination against eggs unlike their own. For them, the cuckoo has not evolved a mimetic egg, but this raises an interesting evolutionary question: how long might it take for egg discrimination to evolve in a host population?

The conclusion is that students interested in behavioural adaptations should clearly not only study function, they should also look at causation, development and evolution. The future trend may well be for behavioural ecologists to rediscover ethology, namely the study and interrelationships between the four questions first asked by Niko Tinbergen (1963).

ACKNOWLEDGEMENTS

I thank Tim Clutton-Brock, Rudi Drent and Walter Koenig for helpful discussion.

REFERENCES

Andersson, M. (1982) Female choice selects for extreme tail length in a widowbird. *Nature, Lond.*, **299**, 818–20.

Brown, J.L. (1987) *Helping and Communal Breeding in Birds*. Princeton University Press, Princeton.

Burke, T., Davies, N.B., Bruford, M.W. and Hatchwell, B.J. (1989) Parental care and mating behaviour of polyandrous dunnocks *Prunella modularis* related to paternity by DNA fingerprinting. *Nature, Lond.*, **338**, 249–51.

Clutton-Brock, T.H. (ed) (1988) *Reproductive Success*. University of Chicago Press, Chicago.

Clutton-Brock, T.H., Guinness, F.E. and Albon, S.D. (1982) *Red Deer: behaviour and ecology of two sexes*. University of Chicago Press, Chicago.

Clutton-Brock, T.H., Albon, S.D. and Guinness, F.E. (1984) Maternal dominance, breeding success and birth sex ratios in red deer. *Nature, Lond.*, **308**, 358–60.

Clutton-Brock, T.H. and Harvey, P.H. (1984) Comparative approaches to investigating adaptation, in *Behavioural ecology: an evolutionary approach*, 2nd edn, (eds J.R. Krebs and N.B. Davies) Blackwell Scientific Publications, Oxford, pp. 7–29.

Cullen, E. (1957) Adaptations in the kittiwake to cliff nesting. *Ibis*, **99**, 275–302.

Davies, N.B. (1990) Dunnocks: cooperation and conflict among males and females in a variable mating system, in *Cooperative breeding in birds*, (eds P.B. Stacey and W.D. Koenig), Cambridge University Press, Cambridge, pp. 457–85.

Davies, N.B. and Brooke, M. de L. (1988) Cuckoos versus reed warblers: adaptations and counteradaptations. *Animal Behaviour*, **36**, 262–84

Davies, N.B. and Brooke, M. de L. (1989a) An experimental study of co-evolution between the cuckoo *Cuculus canorus* and its hosts. I. Host egg discrimination. *Journal of Animal Ecology*, **58**, 207–24.

Davies, N.B. and Brooke, M. de L. (1989b) An expeimental study of co-evolution between the cuckoo *Cuculus canorus* and its hosts. II. Host egg markings, chick discrimination and general discussion. *Journal of Animal Ecology*, **58**, 225–36.

Dijkstra, C., Bult, A., Bijlsma, S. *et al.* (1990) Brood size manipulations in the kestrel (*Falco tinnunculus*): effects on offspring and parent survival. *Journal of Animal Ecology*, **59**, 269–85.

Emlen, S.T. and Wrege, P.H. (1988) The role of kinship in helping decisions among white-fronted bee-eaters. *Behavioural Ecology and Sociobiology*, **23**, 305–15.

Grafen, A. (1988) On the uses of data on lifetime reproductive success, in *Reproductive Success*, (ed T.H. Clutton-Brock), Chicago University Press, Chicago, pp. 454–71.

Hamilton, W.D. (1964) The genetical evolution of social behaviour. I, II. *Journal of Theoretical Biology*, **7**, 1–52.

Jamieson, I.G. (1989) Behavioural heterochrony and the evolution of birds' helping at the nest: an unselected consequence of communal breeding? *The American Naturalist*, **133**, 394–406.

Jamieson, I.G. and Craig, J.L. (1987) Critique of helping behaviour in birds: a departure from functional explanations, in *Perspectives in ethology* Vol. 7., (eds P. Bateson and P. Klopfer), Plenum Press, New York, pp. 79–98.

Koenig, W.D. (1990) Opportunity of parentage and nest destruction in polygynandrous acorn woodpeckers: an experimental study. *Behavioural Ecology*, **1**, 55–61.

Koenig, W.D. and Mumme, R.L. (1987) *Population ecology of the cooperatively breeding acorn woodpecker*. Princeton University Press, Princeton.

Koenig, W.D. and Mumme, R.L. (1990) Levels of analysis and the functional significance of helping behaviour, in *Interpretation and Explanation in the Study of Animal Behaviour*, (eds M. Bekoff and D. Jamieson), Westview Press, Boulder, San Francisco and Oxford, pp. 268–303.

Lack, D. (1965) *The life of the robin*. Witherby, London.

Lack, D. (1966) *Population studies of birds*. Clarendon Press, Oxford.

Lack, D. (1968) *Ecological adaptations for breeding in birds*. Methuen, London.

Ligon, J.D. and Stacey, P.B. (1989) On the significance of helping behaviour in birds. *The Auk*, **106**, 700–5.

Maynard Smith, J. (1982) *Evolution and the theory of games*. Cambridge University Press, Cambridge.

Parker, G.A. (1970) The reproductive behaviour and the nature of sexual selection in *Scatophaga stercoraria* (Diptera: Scatophagidae). II. The fertilization rate and the spatial and temporal relationships of each sex around the site of mating and oviposition. *Journal of Animal Ecology*, **39**, 205–28.
Regelmann, K. and Curio, E. (1986) Why do great tit (*Parus major*) males defend their brood more than females do? *Animal Behaviour*, **34**, 1206–14.
Sherman, P.W. (1989) The clitoris debate and the levels of analysis. *Animal Behaviour*, **37**, 697–8.
Stacey, P.B. and Koenig, W.D. (eds) (1990) *Cooperative breeding in birds*. Cambridge University Press, Cambridge.
Stephens, D.W. and Krebs, J.R. (1986) *Foraging theory*. Princeton University Press, Princeton.
Tinbergen, N. (1953) *The Herring Gull's World*. Collins, London.
Tinbergen, N. (1963) On aims and methods of ethology. *Zeitschrift für Tierpsychologie*, **20**, 410–33.
Tinbergen, N. (1990) *Social behaviour in animals*. Facsimile reprint of 1964 edition. Chapman and Hall, London.
Tinbergen, N. (1973) *The animal in its world : Laboratory experiments and general papers 1932–1972*. George Allen and Unwin, London
Tinbergen, N. (1974) *Curious naturalists*. Penguin, London
Tinbergen, N., Kruuk, H., Paillatte, M. and Stamm, R. (1962) How do black-headed gulls distinguish between eggs and eggshells? *British Birds*, **55**, 120–9.
Tinbergen, N., Broekhuysen, G.J., Feekes, F. *et al.* (1963) Eggshell removal by the black-headed gull *Larus ridibundus* : a behaviour component of camouflage. *Behaviour*, **19**, 74–117.
Trivers, R.L. (1972) Parental investment and sexual selection, in *Sexual selection and the descent of man*, (ed B. Campbell), Aldine, Chicago, pp. 139–79.
Trivers, R.L. (1974) Parent–offspring conflict. *American Zoologist*, **14**, 249–64.
Wilson, E.O. (1975) *Sociobiology : the new synthesis*. Belknap Press, Harvard.

3

From animals to humans

ROBERT A. HINDE

As a DPhil student at Oxford I was doubly fortunate. First, my research supervisor, David Lack, gently accepted that I was not motivated to work on the subject he had intended for me – a comparative study of the feeding ecology of crows – and, while giving me a great deal of help, guidance and stimulation, allowed me to do what I wanted to, a study of behaviour. Second, Niko Tinbergen arrived in Oxford and, not yet firmly established in new research projects, had time to talk with me and teach me. This had a profound influence on me, and has coloured my research ever since. Later we started to write a book together, and one of my greatest regrets is that other demands forced him to forgo authorship: it was a loss both to me and to the enterprise. Re-reading, for the purpose of this essay, some of his papers on the applications of ethology to human behaviour, I find that many ideas that have now become widely accepted, including some that until now I thought were my own, were there already.

The application of a biological approach to human behaviour has met with two major obstacles – biologists and popularizers who so overstated the case that they provoked opposition and derision from social scientists, and social scientists who denied totally the relevance of biological considerations. Tinbergen fell into neither of these groups. In his carefully argued review paper – and I refer especially to the *Science* article 'On war and peace in animals and man' (1968), his Croonian lecture on 'Functional ethology and the human sciences' (1972) – and in his later work on autism (Tinbergen and Tinbergen, 1983), he naturally pressed a biological viewpoint, but always with discretion and humility.

The 'Study of instinct' (1951) contained only 6 pages on the ethological study of man. Acknowledging that it had not yet advanced very far, he devoted most of these pages to an attack on the prejudices that regarded human behaviour as not accessible to ethological methods. In keeping with the Zeitgeist amongst ethologists, this was largely an attack on subjectivism. However, he included a number of specific examples of 'processes usually considered as typical of animals (that) are also found in man' – reflexes, coordination between rhythms, aspects of motivation, 'innate releasing mechanisms' and displacement activities. 'Social behaviour in animals' published in 1953, contained only a few references to comparisons between human and animal behaviour.

However, in his later papers he devoted little space to simple parallels between animal and human behaviour. 'What we ethologists do not want' he wrote in his 1968 paper, 'is uncritical application of our results to man'. There were at least three reasons for this. First, he felt that some authors writing in the 1960s, such as Lorenz and Morris, 'present as knowledge a set of statements which are after all no more than likely guesses' (1968), though in other respects he applauded their efforts to find 'the animal roots of human behaviour'. Second, he argued that most writers who had tried to apply ethology to humans had tried to explain human behaviour by selecting, from the diversity of animals' behaviour, facts to suit their theses. 'Therefore' he writes (1968) 'instead of taking this easy way out, we ought to study man in his own right'. And third, an issue to which we shall return later, he saw that 'both our behaviour and our environment have changed so much since cultural evolution began to gather momentum' (1972) that it is more profitable to apply the 'approach' of biology to the phenomena of human behaviour.

It is evident that, during the 1950s and 1960s, Tinbergen gave much thought to the application of ethological principles to humans: in his later papers on human ethology Tinbergen referred several times to the hours he and his wife had spent child-watching. The application of ethological methods and principles to human behaviour was his primary concern. However, it was his pupil Blurton Jones (1972), no doubt inspired by his supervisor, who pioneered the application of ethological methods to the study of human behaviour. Alongside his empirical work, Blurton Jones provided a masterly summary of a rather extreme ethological position, laying emphasis on the need for objective description in terms of behavioural elements, with more global concepts, such as aggression, anxiety and attachment, being eschewed. This had some important consequences, leading for instance to a distinction in child behaviour between those apparently aggressive acts that form part of rough-and-tumble play and true aggression. Blurton Jones's work facilitated a recrudescence of observational studies of children's behaviour but, in my view (Hinde, 1983), was overcritical of current work in developmental psychology. In laying emphasis on the identification of behaviour elements by 'physical description' (i.e. description in principle capable of reduction to movements of muscles, bones, etc.), the importance of 'description by consquence' and of the use of variable means to obtain a given end or goal-directedness was neglected. Whilst studies of particular movement patterns identified by physical description, such as those of sucking by babies (Gunther, 1955; Prechtl, 1958) smiling (van Hooff, 1972) and other adult facial expressions (Eibl Eibesfeldt, 1972, 1975), have greatly enhanced our understanding of some aspects of human behaviour, an out-and-out molecular approach has clear limitations. Stated baldly, humans are not fish, and descriptive methods that were outstandingly successful in studies of sticklebacks in the 1930s - 1950s, cannot be applied directly to the human case. Interestingly, these studies were barely mentioned by Tinbergen in his later writings on human behaviour. Instead he emphasized other aspects of behaviour that humans share with some other species - aspects which are characterized primarily by flexibility rather than by stereotyped movement patterns, such as exploration.

Nevertheless, Tinbergen did insist that description must precede analysis and/or explanation. 'Intense, long, repeated 'plain' or 'simple' observation, guided by a truly inquiring, not prematurely prejudiced state of mind has to come first', was the Tinbergens' (1983) precept for understanding autism, and they reported that their main clues came from 'gesturing, facial and body expressions, details of where the children go, of their starting or stopping, of the orientation of their bodies or body parts etc.' - in fact, a mixture of physical description and description by consequence aimed at understanding the meaning behind actions. In understanding something like autism, they emphasized that observation and description could be more revealing if coupled with comparison with what is 'normal' (1974).

However, an issue on which Tinbergen laid even more emphasis, and which arises directly from ethology's biological basis, lies in his insistence on the need to distinguish between the four questions of immediate causation, development, evolution and function, and not to neglect any one of them if full understanding is to be achieved (Tinbergen, 1963). In fact the *Science* article 'On war and peace' is much more than its title implies, and provides a brilliant overview of the manner in which asking each of these 'Four Whys' can help in the understanding of behaviour, exemplified by the particular problem of human aggressiveness.

In discussing aggression, Tinbergen wrote of the factors determining aggression in individuals and then, unlike many biologists writing at the time, emphasized that additional factors operated in aggression by animal groups and by human groups, but left open how far the special features of human group aggression were directly derived from an animal heritage of group territoriality. Thus, while he stated in his 1968 *Science* article that he considered that hypothesis 'the most likely one', a few paragraphs later he wrote 'Ethologists *tend* (my italics) to believe that we still carry with us a number of behavioural characteristics of our animal ancestors . . . and that group territorialism is one of those ancestral characters'. The next sentence suggests that in using the word 'tend' he was trying to disassociate himself from some ethologists, for he pointed out that 'cultural evolution, which resulted in the parcelling-out of our living space . . . would, if anything, have tended to enhance group territorialism'. I understand Tinbergen to have meant here that human territorialism as a *simple* remnant of our animal ancestry (as claimed for instance by some writers at that time), was at least an oversimplification.

In writing about group aggression, Tinbergen emphasized how members of human groups unite in the face of an outside danger using, like other species, threat gestures against the enemy and friendly communication with members of the in-group. He also acknowledged the role of leaders in the human case, pointing to the manner in which methods of mass communication could be used to disseminate propaganda which exploited our aggressive tendencies.

While in no way disagreeing with Tinbergen's analysis, more recent work might put the emphasis slightly differently. While Tinbergen emphasized the role of external forces in uniting groups – individuals coming together 'in the face of an outside danger' – more recent work has been concerned with the internal forces that cause individuals to form groups and groups to become differentiated from each other. Social categorization may itself cause a significant bias in favour of same-group members. Individuals may be attracted to others similar (or perceived as similar) to themselves, in part because communication is easier between individuals who perceive the world similarly (Kelly, 1955, 1970), and also because others who hold similar attitudes to oneself help to confirm one's social beliefs (Festinger, 1957; Clore and Byrne, 1974). Furthermore, current work regards the tendencies to exaggerate the differences between in-group and the out-group and to see the in-group as superior, as intrinsic to the process of group formation and differentiation (Tajfel, 1978; Rabbie, 1989). Group formation also introduces many new factors relevant to the incidence of aggression – escalation due to the desire of group members to show off their aggressiveness to their peers, or resulting from anonymity within the group; group values condoning violence; the example of charismatic leaders, and so on.

Unlike many other biologists, Tinbergen did not underestimate the role of cultural factors in human warfare. He emphasized social values that made cowardice despicable, the development of more effective weapons, and the capacity to kill without being exposed to the suffering or appeasement gestures of the victims. It can now be argued that we must also take into account the institutionalization of war. Aggressive propensities play little direct part in the behaviour of soldiers in battle: rather their behaviour is influenced primarily by their rights and duties as incumbents of a particular role in the institution of war (Hinde, 1989, 1991). Aggression comes in, as Tinbergen noted, in that the propaganda helping to stabilize the institution of war plays upon aggressive propensities.

We must remember here not only, as Tinbergen pointed out, that human beings are products of an interaction between their biological propensities and the physical and social environment in which they grow up - that environment including the sociocultural structure of beliefs, values and institutions with their constituent roles and so on - but also that the sociocultural structure is itself a product of human beings. Thus we must come to terms with dialectic relations between human behaviour, interactions, relationships and groups, their physical environment, and the sociocultural structure (Hinde, 1987).

Another aspect of human behaviour concerning which Tinbergen drew lessons from the ethology of other species, was education. Impressed by the learning opportunities provided by play-like activities in animals, he questioned the balance between formal instruction and self-initiated exploratory activities in our present educational system. Joining many educational innovators, he pleaded for less inbibing of 'knowledge' and more 'self-activity', arguing that too much instruction suppresses exploratory learning and sets up resistances against further instruction. He stressed the need for the child to be given security in order to maximize playful exploration, and he discussed at some length (Tinbergen manuscript) the role of adults in children's exploratory play. Acknowledging the need for sensitive participation (cf. Vygotsky, 1934), Tinbergen emphasized also the deleterious effects of too much interference in young children's play, and the deadening effect of too much formal instruction later.

The issue into which Niko Tinbergen and his wife put most of their energies in the last years of their lives was that of childhood autism. They were convinced that the distinction between normal and autistic children is far from sharp, and application of the same methods as they had used for analysing the threat and courtship behaviour of gulls showed that a conflict between hyperanxiety and frustrated sociality were often involved. They suggested that when this conflict becomes severe the child withdraws and socialization is severely hampered. As a result of this, the child fails to learn from social interaction and exploratory behaviour. This view brought the Tinbergens into head-on collision with many psychiatrists, who believed both that autism has organic causes and that genetic factors play a determining role. This debate continues, and I am in no way competent to give a view: while there has been increasing evidence that some type of organic brain dysfunction is involved (Rutter 1988) and that genetic predispositions are of importance in many cases (Rutter, in press), that in no way argues against the views that psychogenic factors are also important, or that there is a spectrum of cases from normal to extreme autism, or that certain forms of treatment may be capable of providing marked improvement in many cases. The recent view that the specific cognitive defect characterizing autism involves the retarded development of a theory of mind (Baron-Cohen, 1989), is compatible with Tinbergen's thesis.

In writing about aggression, education and autism, one issue pervaded Tinbergen's views about the possible contributions of ethology to the understanding of human behaviour - man's influence on his environment. This was not merely a matter of the impact of culture, though his Croonian lecture emphasized the increasing rate of cultural change and the increase in 'adjustabilty' this requires from individuals in each succeeding generation. Tinbergen was concerned also with the growth in the world's human population, the depletion of non-renewable resources and the accumulation of toxic wastes. While such facts are now well enough known, Tinbergen was among the first to realize their importance. His special concern was how they relate to behaviour: 'The cultural evolution is a behavioural evolution, and with it the relationship between what we are doing and what the new environment requires from us' and 'The prevention of possible disadaptation and the creation of a new adaptedness will be a matter of behavioural planning' (1972). Tinbergen's message is of special importance because, while fully aware of man's flexibility and adaptability, he saw that they are limited, and asked whether they are capable of coping with the new environments we are creating. He focused not only on global questions, but on the intimate details of everyday life. In his essay on education he wrote:

> 'Once he had experienced these other types of child-rearing, the contrast with our modern society becomes even more striking. Relatively suddenly, conditions for under-fives have become far less conducive for play. Families are smaller. The work of fathers is done

> more in the past far away from home, and is anyway beyond a child's comprehension. More and more mothers go out to work. Contact between families is reduced. The street, until fairly recently such a widely used playground, has been made unsafe by the motor car. When the parents are at home, they are often tired and irritable, or rushed, or preoccupied. Children see less and less of the craftsman-at-work, and of the mother doing household chores; and anyway the mechanical domestic appliances offer little scope for participation in such chores. What I find even more disturbing is that the mood everywhere around the children has become so serious.

The Tinbergens had lived with the Eskimos, but there is no starry-eyed glamorization of hunter–gatherer in what he wrote. Rather he implied a point-by-point comparison of issues which he believed to be important in child development and asked what consequences the characteristics of the modern world are having on our behaviour. Elsewhere he addressed the relationship between accelerating cultural change, the extended period of human development, and the generation gap. In all these cases he related the demands of modern life to the limitations of the human individual.

I cannot do better, in conclusion, than to quote in full the last three paragraphs of his Croonian lecture, where, after stating the need to identify the new environmental pressures we are creating, he points out the road that must be taken:

> 'And while functional ethology helps us in identifying these pressures, it will be the knowledge of behaviour mechanisms, and of mechanisms of behaviour development, that will have to form the basis for whatever engineering will have to be undertaken.
>
> The execution of such an engineering task may at the moment seem to belong in science fiction, but I am convinced that sooner or later it will become a political issue. Knowing what we do about political decision-making, I believe that it will be useless to call upon people's altruism or use other arguments of a moral nature. Rather, the scientist will have to point out that the prevention of a breakdown, and the building of a new society is a matter of enlightened self-interest, of ensuring survival, health and happiness of the children and grandchildren of all of us – of people we know and love.
>
> No one can say how soon science will be called upon for advice, but if and when that time comes, we shall have to be better prepared than we are now. The main purpose of my paper is therefore to urge all sciences concerned with the biology of Man to work for an

integration of their many and diverse approaches, and to step up the pace of the building of a coherent comprehensive science of Man. In this effort towards integration, animal ethology cannot stand aside – indeed I for one believe that provided it will be given the opportunity for futher development, it can render invaluable services.'

There is a special message here for those interested in conservation. The work of biologists attempting to conserve species or habitats has been less effective than it might have been because of inadequate regard for human behaviour. Physicists and chemists can unravel the nature of atmospheric pollution, but ultimately the cure lies in the hands of economists and social scientists. We should do well to bear Niko Tinbergen's words constantly in mind if we wish to save the world from destruction.

REFERENCES

Baron-Cohen, S. (1989) The autistic child's theory of mind: a case of specific developmental delay. *Journal of Child Psychology and Psychiatry*, **30**, 285–98.

Blurton Jones, N. (1972) *Ethological studies of child behaviour*. Cambridge University Press, Cambridge.

Clore, G.L. and Byrne, D. (1974) A reinforcement-effect model of attraction, in *Foundations of interpersonal attraction*, (ed T.L. Huston) Academic Press, New York.

Eibl Eibesfeldt, I. (1972) Similarities and differences between cultures in expressive movements, in *Non-verbal communication*, (ed R.A. Hinde) Cambridge University Press, Cambridge.

Eibl Eibesfeldt, I. (1975) *Ethology*. Holt, Rinehart & Winston, New York.

Festinger, L. (1957) *A theory of cognitive dissonance*. Rowe, Peterson, Evanston, Illinois.

Gunther, M. (1955) Instinct and the nursing couple. *Lancet*, **1955**, 575–78.

Hinde, R.A. (1983) *Ethology and child development*, Mussen Handbook of Child Psychology, Vol. II (eds M.M. Haith and J. Campos), Jon Wiley, New York pp. 27–94.

Hinde, R.A. (1987) *Individuals, relationships and culture*. Cambridge University Press, Cambridge.

Hinde, R.A. (1989) Towards integrating the behavioural sciences to meet the threats of violence and war. *Medicine & War*, **5**, 5–15.

Hinde, R.A. (1991) (ed.) *The Institution of War*. Macmillan, London.

Hooff, J.A.R.A.M. van (1972) A comparative approach to the phylogeny of laughter and smiling in, *Non-verbal communication*, (ed R.A. Hinde), Cambridge University Press, Cambridge.

Kelly, G.A. (1955) *The psychology of personal contacts*. Norton, New York.

Kelly, G.A. (1970) A brief introduction to personal contact theory, in *Perspectives on personal contact theory*, (ed. D. Bannister), Academic Press, London.

Prechtl, H.F.R. (1958) The directed head-turning response and allied movements of the human body. *Behaviour,* **13**, 212–42.
Rabbie, J.M. (1989) Group processes as stimulant of aggression, in *Aggression and war,* (eds J. Groebel and R.A. Hinde), Cambridge University Press, Cambridge.
Rutter, M. (1988) Biological basis of autism, in *Preventative and curative intervention in mental retardation,* (eds F.J. Mendascino and J.A. Stark), Brookes, Baltimore.
Rutter, M. (in press) Autism as a genetic disorder, in *Advances in psychiatriic genetics,* (eds P. McGuffin and R. Murray), Heinemann, Oxford.
Tajfel, H. (ed) (1978) *Differentiation between social groups.* Academic Press, London.
Tinbergen, N. (1951) *The study of instinct,* Clarendon Press, Oxford.
Tinbergen, N. (1953) *Social behaviour in animals.* Methuen, London.
Tinbergen, N. (1963) On the aims and methods of ethology. *Zeitschrift für Tierpsychologie,* **20**, 410–33.
Tinbergen, N. (1968) On war and peace in animals and man. *Science,* NY **160**, 1411–18.
Tinbergen N. (1972) Functional ethology and the human sciences (Croonian Lecture). *Proc. Royal Society of London,* **B, 182**, 385–410.
Tinbergen, N. (manuscript). *The importance of being playful.*
Tinbergen, N. (1974) Ethology and stress diseases. *Science,* NY **185**, 20–7.
Tinbergen, N. (1983) *'Autistic' children.* Allen & Unwin, London.
Vygotsky, L.S. (1934) *Thought and language.* MIT Press, Cambridge, Mass.

4 War and peace revisited

FELICITY A. HUNTINGFORD

INTRODUCTION

The aim of this paper is to identify the influence of Niko Tinbergen's ideas on the modern study of the mechanisms that control behaviour. To do this, it is necessary first to give an account of Tinbergen's views on the subject, then to see how these ideas were developed in the light of the research that they stimulated (the second wave of ethology), and lastly to identify those aspects of modern theory on the causation of behaviour that can be traced back to Tinbergen's ideas and scholarship. While the first of these steps is easy (because Tinbergen wrote so well) and the second relatively easy (because this topic is already well reviewed), the last has proved surprisingly difficult.

I have chosen to illustrate this paper with reference to aggression, partly because Tinbergen used this behaviour to illustrate many of his ideas about causation and partly because the study of aggression illustrates very clearly the problems of modern research into behavioural mechanisms. The title of the paper refers to an article Tinbergen published in 1968 called 'On war and peace in animals and man' in which he spelled out his views about the causation of aggression, as well as its development and functions.

TINBERGEN'S VIEWS ON THE CAUSATION OF AGGRESSION

Tinbergen's views on the mechanisms that cause animals to fight, as expounded in 'The study of instinct' (1951), concern both the external stimuli and the internal factors involved.

The role of external stimuli

Fighting is triggered by one or a few relatively simple key features of an opponent (the sign stimuli), other factors being ignored. These key features tend to be those that characterize a rival (usually an individual of the same species and sex). The red chest of the breeding male three-spined stickleback is the most famous example (Figure 4.1), of which Tinbergen wrote '. . . the fish reacted essentially to the red and neglected the other characteristics'.

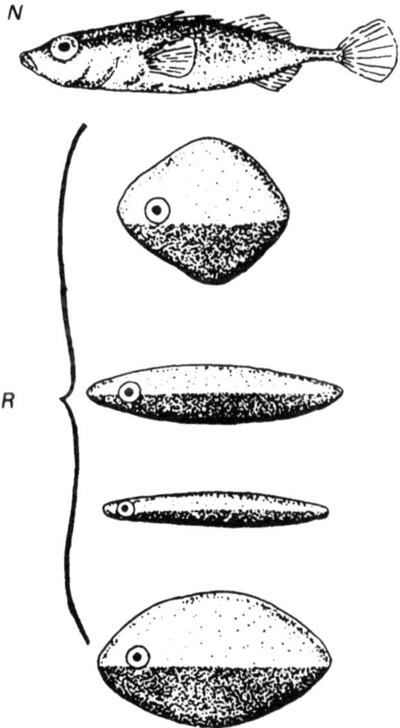

Figure 4.1 The dummies used by Tinbergen in his study of the stimuli that cause aggression in sticklebacks. R indicates dummies with red undersides that were attacked. N indicates an all-silver dummy that was not attacked (from Tinbergen, 1951).

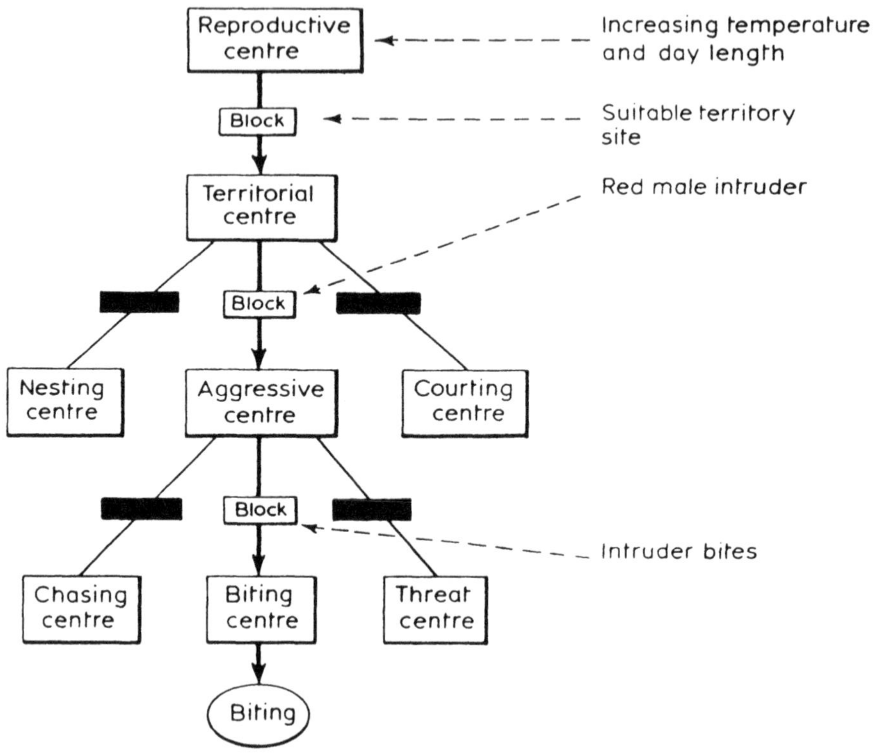

Figure 4.2 Tinbergen's hierarchical model of the control of reproductive behaviour, including aggression, in sticklebacks. Action-specific energy accumulates in and flows between centres at different levels in the hierarchy as sign-stimuli become available; the energy is finally released as the stickleback performs overt behavioural responses (from Huntingford, 1984, after Tinbergen, 1951).

The internal organization of aggression

These key attack-eliciting stimuli interact with a system inside the animal that controls the performance of aggression. Tinbergen depicted this as a hierarchical set of control centres at different levels (Figure 4.2). Considering the stickleback again, when external and internal conditions are right (for example, when day lengths and temperature are increasing in spring) energy accumulates in the highest level reproductive centre. This energy is prevented from flowing to the next level by a block, which is removed by the appropriate sign stimulus (a suitable territory). Once the stickleback is on its territory, energy accumulates in the territorial centre where, once again, descent to the next centre is blocked. The sight of an intruding red male removes this block, so that the aggression centre is activated. Depending on the exact stimulus presented by the rival (whether it bites, threatens or flees), the energy flows down to the next level and so on until the lowest units are activated and the appropriate action is performed. Although this model was based on behavioural data, Tinbergen made use of existing physiological information and clearly saw the centres as anatomically localized neural entities; in other words, it is a software model with strong hardware elements. The model postulates a specific, unitary drive, in that there is just one single system that controls aggression and does not control any other kind of behaviour. In addition, couched as it is in terms of flow between centres and dissipation of drive, Tinbergen's model is clearly based on the concept of drive as an activating or energizing process. One special feature of this famous and widely quoted model is the modest way in which Tinbergen treated it. Thus he writes: 'I should like to emphasize the tentative nature of such an attempt (at synthesis). While such a graphic representation may help us to organize our thoughts, it has grave dangers in that it tends to make us forget its provisional and hypothetical nature', and again, 'It should be emphasized that these diagrams represent no more than a working hypothesis of a type that helps to put our thoughts in order' (Tinbergen, 1951). This is an object lesson of how models should be regarded, especially by their inventors.

Interactions between aggression and other motivational systems

What happens during a fight depends not just on the state of the aggression system but also on a a second system activated by the presence of a rival, namely fear. When an animal is simultaneously motivated to perform aggression and fear responses, these systems interfere with each other's expression. Instead of performing either in its pure form, the animal shows some sort of agonistic display. This view of the causation of displays is usually referred to as the conflict theory. It is interesting to note that Tinbergen explains this dual system of control in terms of the cost of intense, uninhibited fighting, anticipating games theory analyses of animal fights by some 20 years.

According to Tinbergen's conflict theory of the causation of agonistic displays (Figure 4.3), when the level of activation of aggression and fear is low, each is expressed incompletely and the animal shows intention movements or ambivalent postures composed of elements of both. In contrast, when both are strongly activated, there is

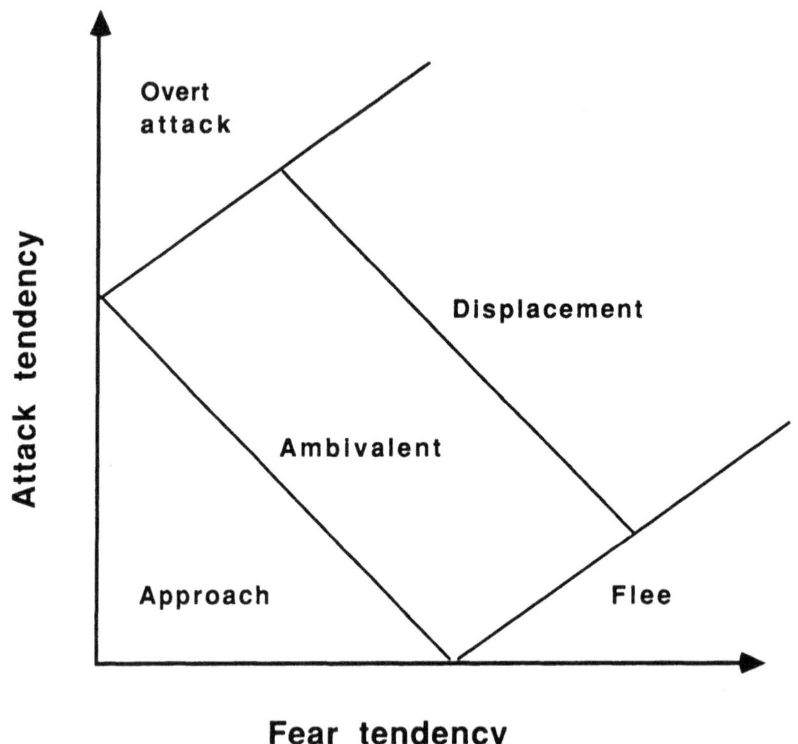

Figure 4.3 Graphical representation of Tinbergen's conflict theory of agonistic behaviour. Actions performed during a fight depend on the level of two independent drives, namely aggression and fear.

complete mutual inhibition. The motivational energy that accumulates in the relevant centre then sparks over into a completely different system (such as nest building) whose activation result in the performance of displacement activities - actions performed out of context that are irrelevant in both functional and causal terms. A classic example is the performance of sand digging (a component of nest building) by male sticklebacks at the territory boundary, where aggression and fear are both strongly activated (Tinbergen, 1951).

THE SECOND WAVE OF ETHOLOGICAL RESEARCH

Tinbergen's writings, with their lucid and critical account of his ideas, helped to stimulate a great deal of behavioural research in the 1960s and this inevitably led to a revision of many of his ideas.

External stimuli

As far as aggression is concerned, it is now clear that the concept of attack as a blind, reflex-like response to one or a few key stimuli is quite wrong. Taking the case of the breeding male stickleback, it is not at all easy to get the fish to respond to models (this in itself casts doubts on their relevance of details such as shape) and when they do respond, red models tend to be attacked less than grey ones (Rowland and Sevenster, 1985), except possibly when the model is very close to the nest (Collias, 1990). In addition, the probability that an intruding male will be attacked depends on many details of the rival - how big it is, whether the resident has met it before and, if so, what happened (Rowland, 1988a) - and on the circumstance of the encounter - whether there is a nest on the territory, whether the nest contains young, and whether there is a predator present (Wootton, 1970; Huntingford, 1977; Ukegbu and Huntingford, 1988). These are all features that determine the costs and benefits of engaging in a fight. Clearly, attack in sticklebacks and in many other animals (Archer, 1987; Huntingford and Turner, 1987) is the result of a complex decision-making process rather than a blind response to a sign stimulus.

The internal organization of aggression

The second wave of ethologists identified a number of weaknesses in Tinbergen's formulation of motivational systems. The problems with unitary drive theories and energy models have been extensively discussed, most especially by Hinde (1960; 1966), so these will be mentioned only briefly here.

Energy models

Analysis of the effects of deprivation and of the factors that bring behavioural sequences to an end do not support the idea that action-specific energy accumulates during quiescent periods and is dissipated by action. In the case of aggression, responsiveness certainly changes in animals that are prevented from fighting, and the performance of aggressive actions influences the internal state of the animals concerned (Archer, 1987; Huntingford and Turner, 1987).

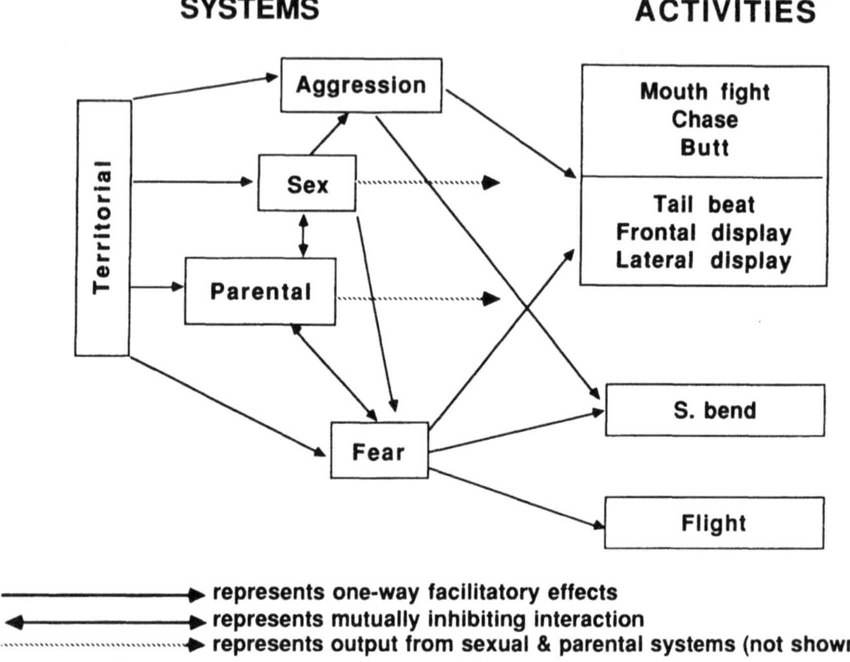

Figure 4.4 A model of the organization of territorial behaviour in a reef fish *Tilapia mariae*, (simplified after Baldicini, 1971).

However, this process is much more complex than the simple build-up and dissipation of energy depicted in Tinbergen's model. A weakness of Tinbergen's motivational model is the one-way flow of information it envisages and its consequent failure to accommodate the various forms of feedback that we now know to occur during fights (and other activities, McFarland, 1971).

Unitary drives

Detailed analysis of the temporal patterning of the various actions shown towards potential rivals often demonstrates the existence of groups of co-occurring behaviour patterns. This justifies postulating a common causal factor influencing performance of all the actions. If such groups of causally-related behaviour patterns all serve a similar overall function, such as deterring or avoiding a rival, then it is legitimate to give these postulated causal factors labels such as aggression and fear, that reflect this function (Figure 4.4).

Interaction between motivational systems
In many cases, the intensity with which an animal attacks and flees are inversely correlated, suggesting that aggression and fear are mutually inhibitory. In addition, some behaviour patterns shown during fights are temporally associated with both aggression and fear (Figure 4.4), so these actions may well be under the joint control of these two systems. However, the conflict theory has been rightly criticized on a number of counts (Andrew, 1972). For example, tendencies other than fear, such as the need to stay put, can interact with aggression to generate displays (Blurton Jones, 1968). In addition, interactions between behavioural systems are now known to be much more complex than the simple inhibitory relationship depicted by the conflict theory. More recent models in the same general framework depict stimulatory and inhibitory links at many different levels and look more like networks than independent hierarchies (Figure 4.5). Tinbergen's model was simple and elegant and there is no virtue in complexity for its own sake; however, these more complex schemes account better for the known behaviour of animals and so represent better explanations of the mechanisms involved.

This increasingly complex picture of the causation of behaviour can be illustrated by the changing explanations of displacement activities Dawkins, 1986). More recent analyses of the situations in which these occur have shown that in spite of the odd appearance of such acts (and we should not forget that they often do look *very* odd), displacement activities are not in fact irrelevant in functional terms; in fighting cocks the opportunity to show displacement pecking during a fight increases the performer's chances of winning. Nor are they irrelevant in causal terms, as the frequency of displacement pecking during fights is influenced by the internal and external factors controlling 'normal' pecking (Feekes, 1972). Thus the sparking-over explanation of displacement activities (already weakened by the failure of energy models to provide satisfactory explanations of the causation of behaviour) gave way to the disinhibition hypothesis. This was first postulated by Andrew (1956) and elaborated by a series of behavioural scientists, including Van Iersel and Bol (1958), Sevenster (1961) and Rowell (1961). According to this view, in an agonistic encounter, a cockerel (for example) may be motivated to feed but this tendency is suppressed both by aggression and by fear. When aggression and fear are aroused simultaneously, the mutual inhibition between them interferes with the ability of both to inhibit pecking, which therefore gains expression; the causal irrelevance of displacement pecking is therefore apparent rather than real. The hypothesis was extended by McFarland (1966), who recognized that disinhibition can occur in a

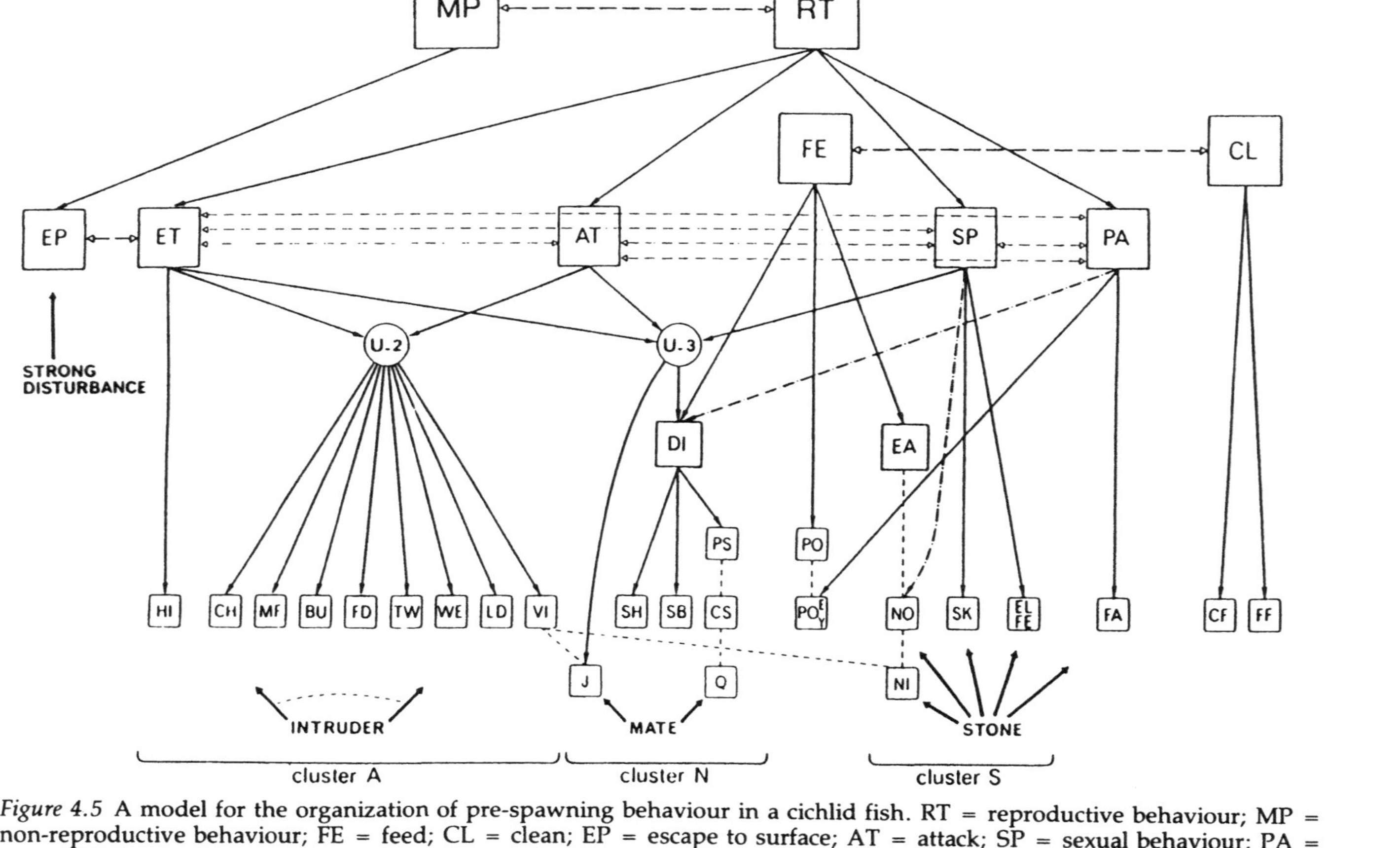

Figure 4.5 A model for the organization of pre-spawning behaviour in a cichlid fish. RT = reproductive behaviour; MP = non-reproductive behaviour; FE = feed; CL = clean; EP = escape to surface; AT = attack; SP = sexual behaviour; PA = parental behaviour; DI = dig; EA = eat algae. u-2 and u-3 are units comparing the relative input of ET versus AT and AT versus SP respectively. All other abbreviations refer to specific behaviour patterns (from Baerends, 1984).

variety of situations that allow animals to switch their attention to other stimuli. The end result of this debate over the causation of displacement activities is much more sophisticated view of the way in which motivational systems interact than Tinbergen gave us, although he was right to recognize that displacement activities can potentially reveal a great deal about behavioural organization.

Tinbergen's influence

In the light of the intense scrutiny that they attracted, it is not surprising that many of Tinbergen's ideas have been altered or even rejected. He himself welcomed these developments and continually revised his theories to accommodate all this new information (for example, in his article 'The aims and methods of ethology' written in 1963), but his role was more important than this, because it is clear from the comments of the key people involved in these debates that Tinbergen actively encouraged them in their work. A glance at the key writing of the time makes this clear, as Tinbergen is warmly acknowledged

Table 4.1 Acknowledgements to Niko Tinbergen

From critics of the lack of feedback in Tinbergen's model
'In particular, I am grateful to Niko Tinbergen . . . for encouraging my initial ventures into the realm of feedback theory.' David McFarland (1971) (Preface) *Feedback Mechanisms in Animal Behaviour.*

From critics of unitary drive concepts and energy models
'I am endebted to a number of colleagues for discussion during the preparation of this paper and especially to . . . Dr N. Tinbergen.' Robert Hinde (1959) Unitary Drives. *Animal Behaviour.*

'I am greatly endebted to [Niko Tinbergen] for his comments [on the manuscript of Animal Behaviour] and also for the many discussions we have had and all I have learned from him'. Robert Hinde (1965) Preface to *Animal Behaviour.*

From critics of Tinbergen's view of displacement activities
'I am grateful to . . . Niko Tinbergen for discussions in the subject in general' Frazer Rowell (1961). Displacement grooming in chaffinches. *Animal Behaviour.*

'The author wishes to express his appreciation to Dr N. Tinbergen for reading the manuscript and for his helpful comments'. David McFarland (1965). Hunger, thirst and displacement pecking in Barbary doves. *Animal Behaviour.*

'Particularly I thank Prof. Dr N. Tinbergen, who introduced me to the study of behaviour and for his kindness and help' Francisca Feekes (1972) 'Irrelevant' ground-pecking in agonistic situations in Burmese jungle fowl.

for his help and encouragement in almost every case (Table 4.1). It is no coincidence that two of the most important books of the time (Hinde's 'Animal behaviour' 1956, and McFarland's 'Feedback mechanisms in animal behaviour' 1971), were dedicated to him. The fertile and encouraging atmosphere that Tinbergen created surely played a critical role in the development of behavioural biology in the 1960s and 1970s. This is an important and valuable legacy, whatever place his ideas may have in modern motivation theory.

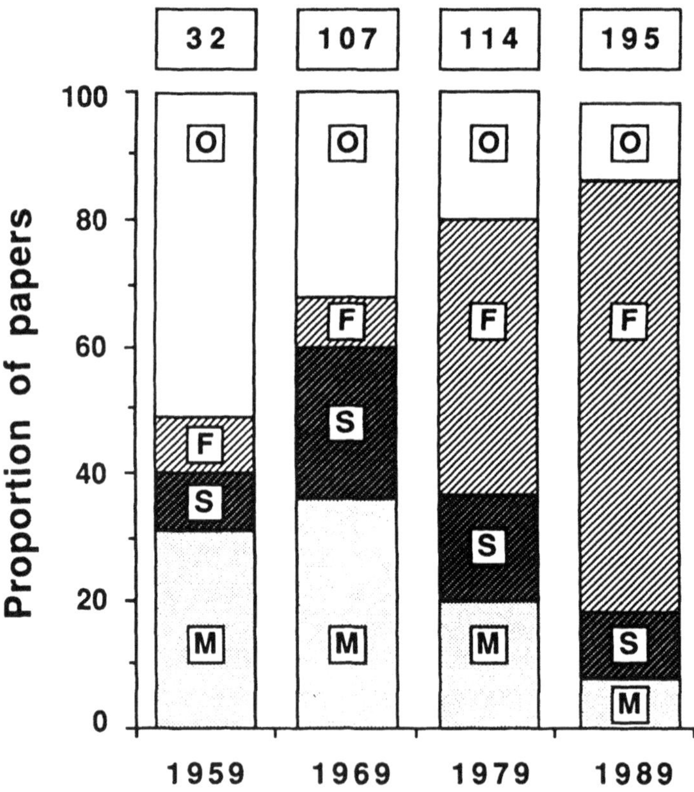

Figure 4.6 The percentage of papers published in three behavioural journals in 1959, 1969, 1979 and 1989 that concerned motivation (M), external stimuli (S), the functions of behaviour (F) and other issues (O; primarily development and genetics of behaviour)

MODERN STUDIES OF CAUSATION

The status of Tinbergen's ideas

So what is the status of Tinbergen's ideas in modern theory on causation? To discover this we need to look not just for modern studies of the causation of behaviour, but for studies of causation by biologists who make use of the behavioural techniques and concepts that he helped to develop. This is not easy, because there is a conspicuous shortage of software studies of the mechanisms of behaviour. Figure 4.6 shows the proportion of papers on different aspects of behaviour published in the three main behavioural journals (*Animal Behaviour, Behaviour* and *Ethology*) in the period 1950 - 1990. The figures confirm what is already well known, namely that in the 1950s and 1960s most people were studying causation and development, but that by the end of the 1980s most people were studying function. The turning point came in the late 1970s; for example, in 1979 Caryl reanalysed a number of classical ethological studies and showed that the data were better explained by game theory models (which predicted that fighting animals should evolve poker faces - Maynard-Smith and Price, 1973) than by ethological models (which predicted that animals should wear their hearts on their sleeves). In the same year, Jakobsson *et al.* used sequence analyses of behavioural exchanges during fights between cichlid fish (in more or less the classical ethological mode) to test and support the same games theory prediction. These two studies represent a clear change of emphasis in research on aggression and other aspects of behaviour.

In spite of this change of emphasis, we can see Tinbergen's influence in modern studies of causation, but this influence is complex and shows up in different ways. Over and above providing a model of how to conduct and write about scientific research, Tinbergen's legacy has several components (Table 4.2).

Table 4.2 The components of the Tinbergen Legacy

The direct legacy: modern studies that incorporate classical ethological techniques and ideas
The slightly-less direct legacy: modern theories that grew out of criticisms of Tinbergen's model of motivation
The indirect legacy: flourishing interdisciplinary studies encouraged by Tinbergen, such as neuroethology and behavioural endocrinology
The (almost) unclaimed legacy: call for links between studies of causation and function.

The direct legacy

There are still biologists who conduct detailed analyses of behavioural sequences in the classic ethological mode, but who go beyond this to ask questions about other aspects of behaviour. Appropriately, much of this is being conducted by Dutch ethologists, particularly Baerends and his colleagues. Their studies, mainly on cichlid fish and birds (Baerends, 1984), draw on Tinbergen's ideas (for example, their models include a dual system for the control of agonistic behaviour (Figure 4.5)), but they also incorporate other more recent developments in behavioural theory. This combination has produced a great deal of information about the mechanisms that control behaviour, and about its development and functions.

To give just one of many possible examples, Groothius (1989) used classic ethological techniques to analyse the causal structure of behaviour in gull chicks, from hatching to adulthood. Among other things, he found that the frequency with which displays such as choking are performed during development correlates positively with the frequency of performance of overt aggression and fear; this is clearly consistent with the conflict hypothesis. The adult form of the choking display develops by the gradual addition of aggressive elements (downwards pointed bill, raised carpels and tilted body) onto the crouching posture typically shown by a frightened chick when hiding in cover (Figure 4.7). Groothius' study identified a key role for the hormone testosterone in the development of agonistic displays in young gulls of both sexes. Levels of circulating testosterone are naturally high at the time when full displays are first shown, and

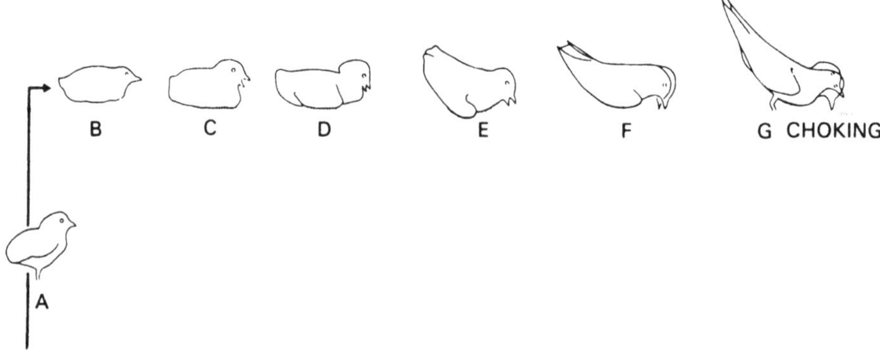

Figure 4.7 Development of choking in the black-headed gull from the crouching response of very young chicks (simplified from Groothius, 1990)

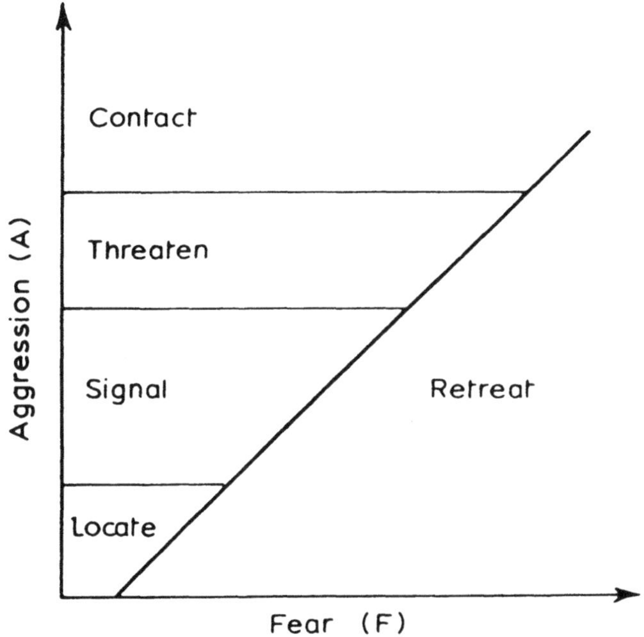

Figure 4.8 A two-factor model of spider fights, in which choice of action depends on the level of independent tendencies to attack or to flee (from Huntingford and Turner, 1987, simplified from Maynard-Smith and Reichert, 1985).

implants of testosterone (which increase aggressiveness) induce the performance of complete displays even in very young gulls. These observations suggest that the role of testosterone is to create a motivational state that permits the development of agonistic displays.

Two recent beneficiaries of the Tinbergen legacy are Maynard-Smith and Riechert (1985) who developed a two-factor model of the control of aggression in spiders (Figure 4.8). This model successfully reconstructs the strategic decisions that spiders are known to take during fights, and gains some support from a genetic analysis that suggests separate patterns of inheritance for aggression (at sex-linked loci) and fear (at autosomal loci) (Riechert and Maynard-Smith, 1989).

The slightly-less-direct legacy

In the examples given above, Tinbergen's ideas and concepts are still proving useful (albeit in a modified form) in interpreting behavioural data. Other important areas of modern motivational theory developed,

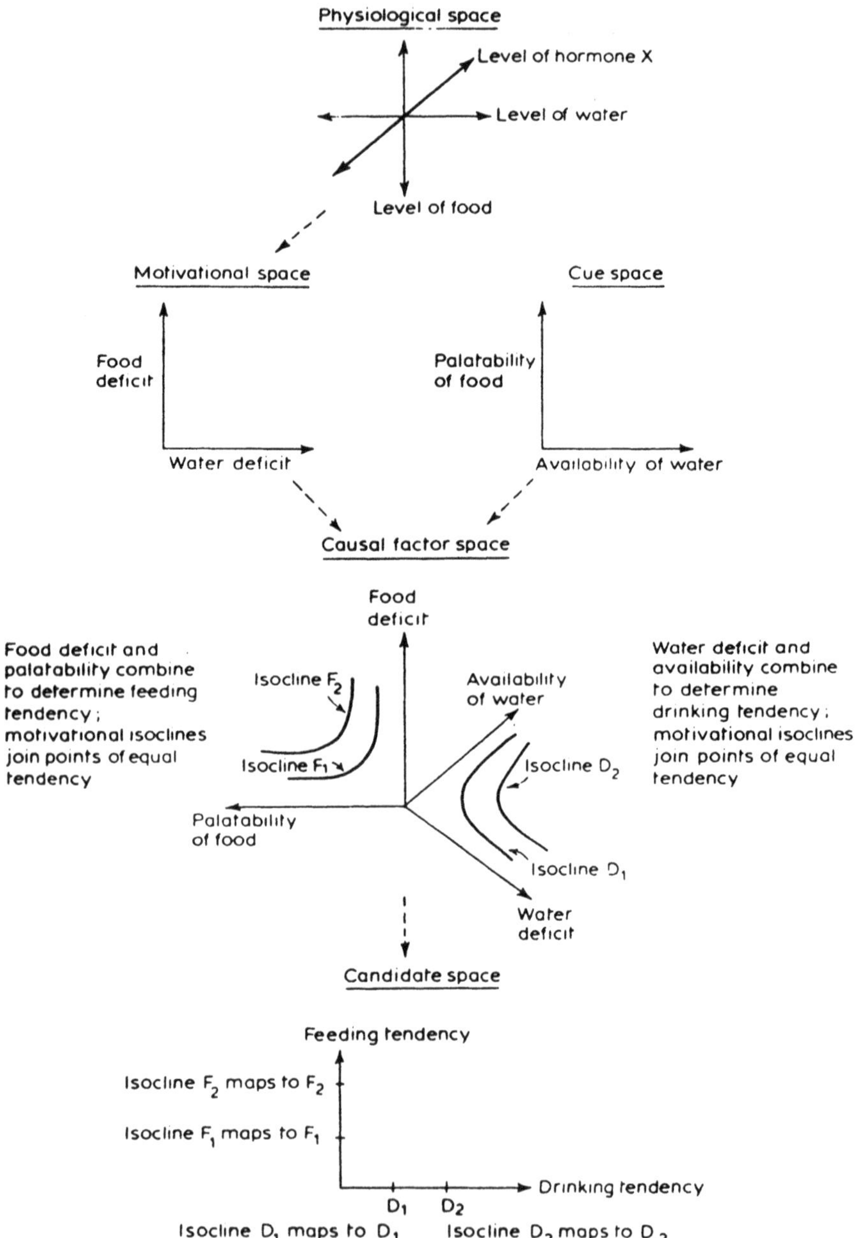

Figure 4.9 McFarland's space-state model of motivation (from Huntingford 1984, after McFarland and Houston, 1981).

in part at least, out of the recognition of the inadequacies of Tinbergen's formulation that became apparent during the second wave of ethology. Because Tinbergen encouraged these studies and welcomed their results, these too are part of his legacy. McFarland's Space–State model of motivation (McFarland and Houston, 1981, Figure 4.9) looks very different from Tinbergen's Hierarchical model [McFarland writes 'Traditionalists in the field of motivation will recognize little in this book that is borrowed or derived from their work' (McFarland, 1974)], but it is still a case in point. In this model, the complex, non-unitary nature of motivational systems is accommodated by making the various internal and external factors that influence behaviour into axes in a multidimensional causal factor space. The various kinds of feedback are handled by depicting the consequences of behaviour as movement through state space. This framework provides a way of characterizing the complexities of behavioural organization and (if one is clever enough) of understanding them.

The indirect legacy

Tinbergen's influence on modern motivation theory can also be seen in the effects of his writing and personality on people working in other disciplines. In particular he contributed to an intellectual climate in which interdisciplinary research could flourish. For example, he actively encouraged the integration of behavioural and physiological studies, as indicated by the following statement: '. . . future work could only be done by workers who are fully acquainted with the instinctive behaviour as a whole and with its analysis, and at the same time are in command of neurophysiological methods and techniques. . . . It is an urgent task of ethologists and neurophysiologists to join efforts in the training of "ethophysiologists"' (Tinbergen, 1951).

Causation and function

Tinbergen stressed the importance of research into the functions of behaviour and one of the reasons for the change in emphasis from studies of causation to studies of function is that he argued very persuasively. It should be said that this extreme trend is not one that he would have welcomed unreservedly, since he also argued for continual interaction between causal and functional studies: '. . . the two approaches are mutually inspiring . . . The discovery of each particular achievement inspires one to find out "how it is done"; conversely, the student of mechanisms derives satisfaction from understanding how the achievements of the mechanisms contribute to the animal's success.' (Tinbergen, 1972). Until recently, this call for links between causal and functional studies has gone largely unheard – it remains an (almost) unclaimed legacy.

It is fitting that one of the few people who have worked actively to find a way of integrating causal and functional studies is Tinbergen's successor at Oxford, David McFarland. In a series of articles and books (1977, 1989), McFarland (together with Alasdair Houston, 1981) has developed a framework that directly links the multivariate mechanisms controlling behaviour to the consequences for fitness of different behavioural options. His aim is articulated most clearly when he asks 'How are animals so organized that they are motivated to do what they ought to do at a particular time?' (McFarland, 1989). His framework is necessarily complex and is not easy to apply, but it highlights important behavioural questions. Both causal and functional studies suffer if developments in these two fields get out of step. It is easy to see how thinking about functions can help studies of causation; for example, if he had known what we now know about the strategic decisions animals make during fights, Tinbergen would not have been satisfied with a theory of causation that had animals driven by an irresistible drive to fight regardless of the consequences, and had them responding blindly to just one feature of a potential rival. However, in the present climate of opinion, it is the opposite case that one has to argue.

I suggest that when they turn their backs on motivational studies, behavioural ecologists deny themselves an important source of information. For example, it is now clear that many animals use simple decision rules to achieve approximately optimal solutions. Recognition of this fact (which in effect concerns the mechanisms that control behaviour) represents an important development in understanding how selection acts on behaviour. A similar rapprochement between functional and causal studies might also be useful in aggression research. To give one example: on the one hand, a long series of games theory models, running from Maynard-Smith and Parker's initial work on assessment (1974) to the recent extension of Enquist's Sequential Assessment Game (Enquist *et al.*, 1990), has interpreted behavioural exchanges during fights as a process of acquisition of accurate information about the relative fighting ability of the contestants. This extremely fruitful approach has helped to answer many difficult questions about why animals fight in the way they do. It may be that from a functional perspective there is no need to know about the mechanisms by which this is achieved. On the other hand, many motivational studies suggest that other important things are going on during fights; a number of studies (Archer, 1987; Huntingford and Turner, 1987) have shown that decisions depend critically on the animal's motivational state, on the current level of aggression of its rival, on whether this is increasing or decreasing and on the part that the animal's own previous actions have played in bringing about such changes. This seems more complicated than is necessary simply to allow animals to assess their relative fighting ability, suggesting that other functional issues are involved. In 1953 Tinbergen wrote '. . . the result of a hostile clash is not determined by the actual strengths of the two contestants but by their "dash", by the strength of their fighting drive' (Tinbergen, 1953). We now know that he was completely wrong to dismiss strength as a determinant of the outcome of fights, but perhaps we should give some thought to the functional significant of 'dash'.

THE TINBERGEN LEGACY

What Niko Tinbergen wrote about the causation of behaviour 20 years ago and more, still has an influence on the way behavioural scientists work. In the first place, some of his ideas (especially his recognition of the interacting effects of aggression and fear) still have direct value as explanatory concepts helping us to study and understand how behaviour is organized. Secondly, by encouraging critical evaluation of his own ideas, he promoted the development of the more sophisticated theories of motivation currently in use. Lastly, by his intellectual broadmindedness he contributed to the intellectual climate in which interdisciplinary research into causation flourished. These are no mean achievements and as beneficiaries of Tinbergen's intellectual legacy we are richly endowed.

ACKNOWLEDGEMENTS

I would like to thank a number of colleagues, including Mike Hansell, Neil Metcalfe and Pat Monaghan, for comments on earlier versions of this paper, and in particular, Neil Metcalfe for last-minute help in preparing the figures for the talk on which it is based.

REFERENCES

Andrew, R.J. (1956) Some remarks on behaviour in conflict situations with special reference to *Emberiza spp. British Journal of Animal Behaviour*, **4**, 41–45.

Andrew, R.J. (1972) The information potentially available in mammalian displays, in *Non-verbal communication*, (ed R.A. Hinde), Cambridge University Press, Cambridge, pp. 179–206

Archer, J. (1987) *The behavioural biology of aggression*, Cambridge University Press, Cambridge.
Baerends, G. (1984) The organization of the pre-spawning behaviour in the cichlid fish *Aequideus portalegrensis*. *Neth. S. Zod*, **34**, 233–366.
Blurton Jones, N.G. (1968) Observation and experiments on the causation of threat displays in the great tit *Parus major*. *Animal Behaviour Monographs*, **1**, 75–158.
Caryl, P.G. (1979.) Communication by agonistic displays: what can games theory contribute to ethology? *Behaviour*, **68**, 136–69.
Collias, N.R. (1990) Statistical evidence for aggressive responses to red by male three-spined sticklebacks. *Animal Behaviour*, **39**, 401–3.
Dawkins, M.S. (1986) *Unravelling animal behaviour*. Longman, Harlow, Essex.
Enquist, M. Leimar, O, Ljungberg, T. *et al.* (1990) A test of the sequential assessment game: fighting in the cichlid fish *Nannacara anomala*. *Animal Behaviour* **40**, 1–14.
Feekes, F. (1972) 'Irrelevant' ground-pecking in agonistic situations in Burmese red jungle fowl (*Gallus gallus spadiceus*). *Behaviour*, **43**, 186–326.
Groothius, T. (1989) On the ontogeny of display behaviour in the black-headed gull, *Larus ridibundus*. PhD Thesis, University of Groningen.
Hinde, R.A. (1960) Energy models of motivation. *Symposium Society for Experimental Biology*, **14**, 199–213.
Hinde, R.A. (1966) *Animal behaviour*, McGraw-Hill, New York.
Huntingford, F.A. (1977) Inter- and intra-specific aggression in male three-spined sticklebacks. *Copeia*, 158–9.
Huntingford, F.A. (1984) *The study of animal behaviour*. Chapman & Hall, London.
Huntingford, F.A. and Turner, A.K. (1987) *Animal conflict*. Chapman & Hall, London.
Jakobsson, S, Radesater, T. and Jarvi, T. (1979) On the fighting behaviour of *Nannacara anomala* (Pisces, cichlidae). *Zeitschrift für Tierpsychologie*, **49**, 210–20.
McFarland, D.J. (1966) The role of attention in the disinhibition of displacement activities. *Quarterly Journal of Experimental Psychology*, **18**, 19–30.
McFarland, D.J. (1971) *Feedback mechanisms in animal behaviour*, Academic Press, London.
McFarland, D.J. (1974) *Motivational control systems analysis*, Academic Press, London.
McFarland, D.J. (1977) Decision-making in animals. *Nature*, **269**, 15–21.
McFarland, D.J. (1989) *Problems of animal behaviour*, Longman Scientific, Harlow.
McFarland, D.J. and Houston, A.I. (1981) *Quantitative ethology*. Pitman, London.
Maynard-Smith, J. and Parker, G. (1974) The logic of asymmetric contests. *Animal Behaviour*, **24**, 159–75.
Maynard-Smith, J. and Price, G. (1973) The logic of animal conflict. *Nature*, **246**, 15–18.
Maynard-Smith, J. and Riechert, S.E. (1985) A conflicting-tendency model of spider agonistic behaviour: hybrid–pure line comparisons. *Animal Behaviour*, **32**, 564–78.
Riechert, S.E. and Maynard-Smith, J. (1989). Genetic analysis of two behavioural traits linked to individual fitness in the desert spider,

Agelenopsis aperta. *Animal Behaviour*, **37**, 624–37.
Rowell, C.H.F. (1961) Displacement grooming in the chaffinch. *Animal Behaviour*, **9**, 38–63.
Rowland, W.J. (1988a) The effect of body size, aggression and nuptial colouring in competition for territories in male three-spined sticklebacks, *Gasterosteus aculeatus*. *Animal Behaviour*, **36**, 629–38.
Rowland, W.J. (1988b) Aggression versus courtship in three-spined sticklebacks and the role of habituation to neighbours. *Animal Behaviour*, **34**, 348–57.
Rowland, W.J. and Sevenster, P. (1985) Sign stimuli in the three-spined stickleback (*Gasterosteus aculeatus*): a reexamination and extension of some classic experiments. *Behaviour*, **93**, 241–57.
Sevenster, P. (1961) A causal analysis of a displacement activity (fanning in *Gasterosteus aculeatus*). *Behaviour*, (Suppl.) **9**, 1–170.
Tinbergen, N. (1951) *The study of instinct*. Clarendon Press, Oxford.
Tinbergen, N. (1953) Fighting and threat in animals. *New Biology*, **14**, 9–23.
Tinbergen, N. (1963) On aims and methods of ethology. *Zeitschrift für Tierpsychologie*, **20**, 410–33.
Tinbergen, N. (1968) On war and peace in animals and man. *Science*, **160**, 1411–18.
Tinbergen, N. (1972) Functional ethology and the human sciences. *Proceedings of the Royal Society*, **B 182**, 385–410.
Ukegbu, A.A. and Huntingford, F.A. (1988) Brood value and life expectancy as determinants of parental investment in male three-spined sticklebacks, *Gasterosteus aculeatus*. *Ethology*, **78**, 72–82.
Wootton, R.J. (1970) Aggression in the early phases of the reproductive cycle of the male three-spined stickleback (*Gasterosteus aculeatus*). *Animal Behaviour*, **18**, 740–46.

5

Animal communication: ideas derived from Tinbergen's activities

JOHN R. KREBS

INTRODUCTION

When I was a new research student in the late 1960s, four things struck me about Niko Tinbergen. The first was the breadth of his approach. I personally benefited greatly from this because I wanted to study a problem in population ecology and I ended up in his Animal Behaviour Group. Niko quite rightly saw no real distinction between the study of behaviour and the study of ecology.

The second thing that struck me was his insistence on great precision of thought. I can remember one of the first Behaviour Group seminars I ever went to: there was an invited speaker from outside Oxford who stood up and managed to get through two sentences before the discussion broke down into exactly what he meant by those two sentences. The speaker went home without ever having finished the rest of his seminar.

A third characteristic of Niko that has been referred to by many other people was his total lack of hierarchy and pomp. This made a particular impression on me because he was at that time at the height of his fame and a well-known international figure, but I never felt, in talking to him, anything other than a colleague and an equal, one whose views were to be judged and appreciated in the same way as those of more senior colleagues.

Finally, there was a feature of Niko that made a great impression on me as an undergraduate, namely, his athletic skills. During one of his undergraduate lectures, he noticed that someone at the back of the steeply tiered lecture theatre that we used in the old zoology department was reading a newspaper. Without even hesitating, and continuing to give his lecture, Niko leapt on to the front bench, strode two benches at a time up to the top row, grabbed the newspaper scrunched it up into a ball and then strode back down to the front as though nothing had happened.

Turning now to the theme of communication, I have to confess that initially I thought it was going to be relatively easy to define Niko's contribution to modern studies of animal communication. Many of his views are contained in his classic paper on 'derived activities' (1952). I had thought that what I would do would be to take, say, five major ideas from that paper, then five key ideas from current literature and trace the threads between them. It turned out, however, to be much more difficult than I had thought. The views of communication that were prevalent in 1952 and even up to the time of the Symposium on Ritualization organized by Huxley in 1966 are very different from present-day ideas, and the connections between then and now are not at all straightforward. So what I will do is summarize from a purely historical point of view what Tinbergen said in his 1952 paper and say something about the ideas on communication that were around at the time. I will try to give a picture as I see it, of what might be called the Tinbergen/Lorenz view, following this with what I perceive to be the important current issues in the field of animal communication, and the links between them, and then pick out possible future directions for research.

THE TINBERGEN/LORENZ VIEW OF COMMUNICATION

A major part of Tinbergen's (1952) view of communication was to do with theory of 'derived' activities – that is, the idea that what we now see as signals were derived in the course of evolution from other non-signal movements, particularly those shown by animals in situations of motivational conflict. Examples of behaviour that Tinbergen saw as the 'raw material' for evolution into signals were intention movements, ambivalent behaviour, protective responses (a later suggestion of Richard Andrews), autonomic responses, displacement activities and redirected activities (Fig. 5.1). All of these were thought to arise when an animal was motivated to do two incompatible things at once (for example, to attack and to flee). Tinbergen saw this as the activation of two specific 'drives' and one of the issues he addressed was whether, as signals evolved, they became emancipated or freed from their original causal factors. Signals may have evolved from situations of motivational conflict but by the time they had taken on a communication function, the original conflict may have been lost.

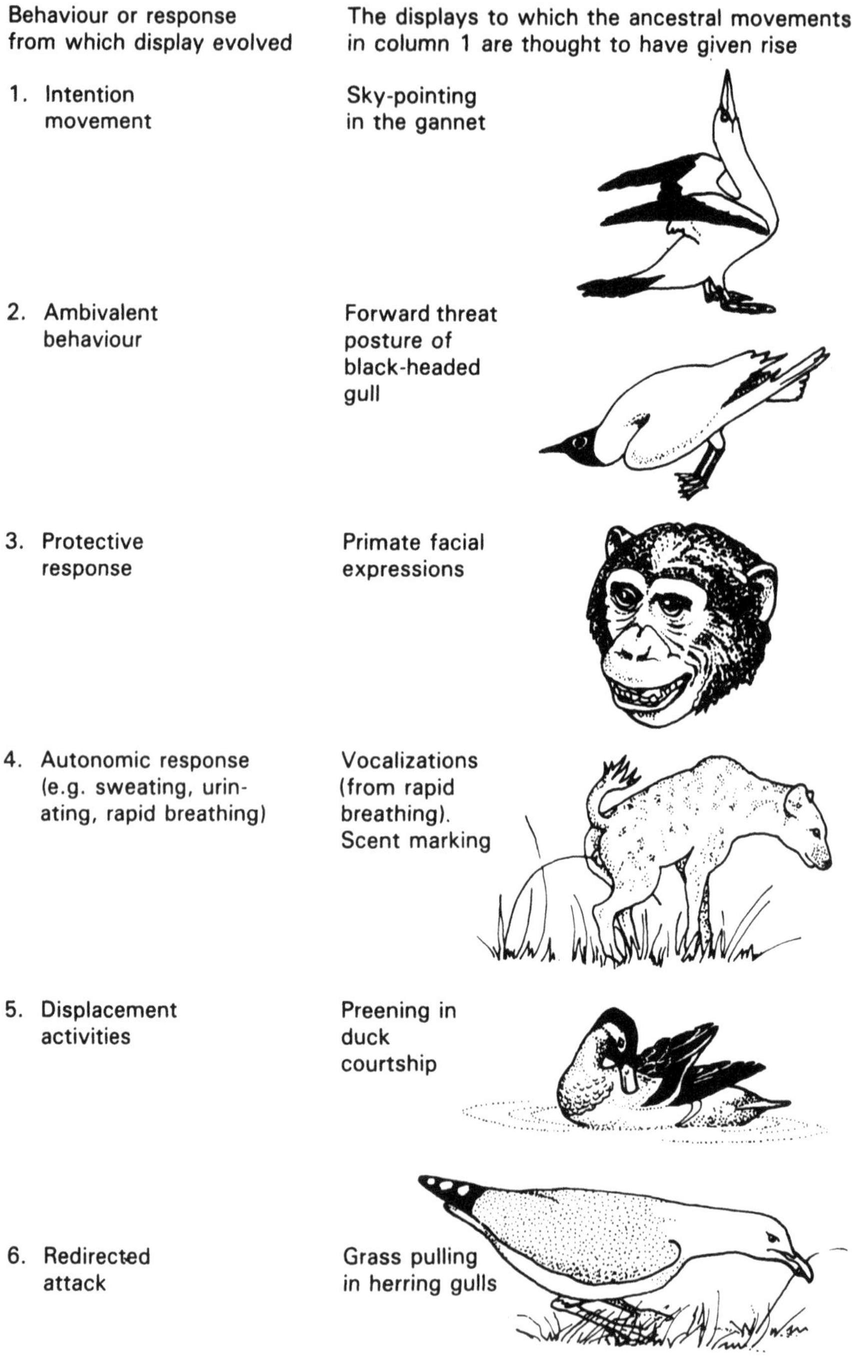

Behaviour or response from which display evolved	The displays to which the ancestral movements in column 1 are thought to have given rise
1. Intention movement	Sky-pointing in the gannet
2. Ambivalent behaviour	Forward threat posture of black-headed gull
3. Protective response	Primate facial expressions
4. Autonomic response (e.g. sweating, urinating, rapid breathing)	Vocalizations (from rapid breathing). Scent marking
5. Displacement activities	Preening in duck courtship
6. Redirected attack	Grass pulling in herring gulls

Figure 5.1 Examples of the kind of behaviour pattern and other responses from which displays in birds, fish and primates are thought to have evolved. (After Hinde (1970); redrawn from Krebs and Davies (1987) *An introduction to behavioural ecology*. Blackwell Scientific Publications, Oxford.)

Tinbergen was, therefore, concerned with the motivation or causal basis of signal movements.

A second aim of Tinbergen's theory was to describe and then to account for the changes that signals underwent as they evolved from ancestral conflict behaviour to fully formed signals. This is the process referred to as 'ritualization', in which signals become exaggerated, stereotyped and repeated. The evidence for the idea that signals did change in this way came from comparative studies of closely related species. The ancestral ground-pecking movements of the jungle fowl courting a hen, for example, seem to have evolved into the exaggerated movements of the peacock in which no real food is involved. Two 'snapshots' of modern species give a picture of how evolutionary change might have taken place in the past. Tinbergen argued that ritualization increases signal effectiveness. Cullen (1966) made this idea rather more explicit by arguing that ritualized signals evolve because they reduce ambiguity. With a ritualized signal, an animal can signal clearly 'I am going to attack you' and not 'I am going to flee' or 'I want to mate' not 'I am not interested in mating'.

A third aspect of Tinbergen's theory was that as well as evolutionary history, we need to know about the present-day function of signals. Then, as now, ethologists tended to concentrate on courtship and threat signals, mainly in a restricted range of animals. Tinbergen saw courtship signals as serving functions such as attracting a mate, stimulating or arousing a partner, synchronizing the sexual activity of a pair, or cementing a pair bond. He saw threat signals as serving the function of conveying the signaller's intentions about how likely he was to attack a rival.

Tinbergen was also concerned to give an account of the variety and design of signals. This means he wanted to understand why signals have the structure they do, why the same animal uses different signals at different times and why there are such differences between species in what signals they use. He argued that one reason for such differences was differences in the 'raw material' from which the signals evolved, which in turn was a reflection of the particular conflict behaviour from which they derived. For example, if a signal was derived from displacement preening, as seems to be the case in certain species of duck, then this would affect what the final ritualized signal looked like. One of the arguments he used to explain why the same animal may use different signals on different occasions was that the signal used may depend on the posture it happens to be in at the moment of signalling. If the animal happened to be sitting down it might use signal A, and if it happened to be standing up it might use signal B. As well as the posture of the animal itself, external stimuli such as whether there was food present, might affect exactly which signal the animal used. His account of the design and variety of signals was, therefore, essentially a causal one.

THE MODERN VIEW OF ANIMAL COMMUNICATION

I now come to the modern view of animal communication. There has been surprisingly little work on the causal basis of displays over the last 25–30 years, as Huntingford (Chapter 4) also points out, and so I shall say very little about this, concentrating instead on those aspects that have attracted considerable attention – namely, evolutionary change, function and variety and design.

I will deal first with theories about evolutionary change. Here we find two recent views that are decidedly different from the evolutionary ritualization of signals that Tinbergen described. One of these views was put forward by Richard Dawkins and myself (1978; also Krebs and Dawkins, 1984). The other has been advocated in a series of articles by Zahavi (1975; 1979; 1987). Although these two appear to be very different, I believe that they are essentially similar and are best seen as reflecting two sides of the same hypothesis. They both differ from Tinbergen's account of animal signalling by emphasizing the role of *coevolution* between actors (the ones performing the signals) and reactors (the ones responding to them). The key point is that the interests of actors and those of reactors are not the same. The 'selfish gene' view of animal behaviour (Hamilton, 1964; Dawkins, 1976) sees the coevolution between actors and reactors as an 'arms race' in which actors are selected to increase the effectiveness of their signals and to change the behaviour of reactors to their own ends, while reactors will be selected to increase their discrimination and their 'sales resistance'. Both the theories of Dawkins and Krebs (1978) and that of Zahavi (1975; 1989) are similar in emphasizing the selfishness of the two parties involved. Both see ritualization as a produce of a coevolutionary arms race. Where they appear to differ is in whether the reactors can be described as being 'manipulated' by the salesmanship of the actors.

Dawkins and Krebs (1978; also Krebs and Dawkins, 1984) considered cases where reactors appear to get the worst of a coevolutionary arms race and to be manipulated into doing things that are not good for them. Two obvious examples are reed warblers being 'manipulated' by cuckoos, and firefly females attracting and then eating males of another species by emitting their species-specific pattern of flashes (Lloyd, 1979). One possible explanation for such cases of apparent exploitation is frequency dependence. If cuckoos in the nest or predators emitting courtship signals are relatively rare events compared to the real thing, then it may on average pay a reactor to respond, even if occasionally it ends in disaster.

However, there is a more general reason why reactors should be vulnerable to exploitation by actors and this can be traced right back to Tinbergen's view of how signals evolve in the first place. Tinbergen (1952) argued that the reason why conflict behaviour so often provides the raw material for signal evolution is that when an animal is in a conflict, it is in a state of transition between two motivational states. The behaviour that it performs in such a situation therefore has a high predictive value about what it will do next. The reactors can therefore 'mind read' or predict the behaviour of the actor (Krebs and Dawkins, 1984) and this in turn makes them vulnerable to exploitation over evolutionary time by actors which then change their behaviour, so as to manipulate reactors.

Krebs and Dawkins (1984) also made a distinction between two kinds of signal which would have different paths of coevolution. With non-cooperative signals (most of the signals studied by ethologists) the path would lead to an arms race of persuasion by the actor and sales resistance by the reactor, and consequently to the evolution of conspicuous, repetitive advertising signals which bear all the hallmarks of 'ritualization'. Other signals, however, which can be called cooperative because they occur between relatives or between members of a mated pair, will be selected to be the opposite of ritualized. For reasons of economy of production, they will evolve towards diminished amplitude, diminished conspicuousness and heightened receiver sensitivity. If it pays the reactor to receive and respond to a signal, it will be straining its ears or eyes, so that the actor has no need to produce a loud blast of sound or bright colours. Because of their inconspicuous nature, ethologists may have overlooked many examples of cooperative signals. Our point was that not all signals should be ritualized and we suggested that a possible line of research would be to see whether cooperative signals are on the whole the hushed conspiratorial whispers we predicted, with non-cooperative signals the bright, noisy conspicuous signals that Tinbergen studied.

In apparent contrast to the Dawkins–Krebs view of communication, which sees many instances of animals 'manipulating' each other, Zahavi (1975, 1989) emphasizes the essential honesty of signals. He argues that reactors should never respond to a signal – such as a threat or courtship display – unless it is an honest indictor of something the reactor was interested in, such as fighting ability or suitability as a mate. For example, if females benefit from choosing disease-free males (Hamilton and Zuk, 1982), then they should be selected to respond to some feature of a male that gives a genuine indication of his disease-resistance. Unless signals are honest, reactors will not respond to them. The only way in which honesty can be maintained is if signals are costly and only those signals which are, say, genuinely healthy and free of disease will be able to pay the cost. Differences between individuals in their ability to pay such costs (determined ultimately by how healthy and disease-resistant they are) explain why individuals also differ in their level of signalling.

Zahavi further argued that in the design of signals there should be a relationship between cost (indication of quality that the signal is concerned with) and the structure of the signal. Several examples of signalling appear to show this relationship quite clearly. Fitzgibbon and Fanshawe (1988) studied stotting in gazelles, a curious behaviour in which the animals leap into the air while fleeing from a predator. They showed that gazelles that stot at a high rate are more likely to escape when attacked by hunting dogs than those that stot at a low rate. Stotting does therefore seem to be a genuine indication of escape ability. The hunting dogs react to it by picking selectively on gazelles that stot at a low rate. In the dry season, when conditions are poor, the gazelles do not stot very much, as they are not in good enough condition to do so.

Borgia (1985) provides another example of 'honest' signalling, this time in the courtship display of bower birds. The male satin bower bird (*Ptilinorynchus violaceus*) in common with other closely related species, decorates its bower with artefacts, sometimes natural, sometimes unnatural. Females choose as mates those males with the largest number of artefacts in their bowers. Now the number of artefacts a male bower bird has is an honest indication of his fighting ability, because one of the main ways the males have of collecting artefacts is to steal them from other males. So the males that have a lot of ornaments in their bowers are inevitably those that are best at stealing from others and also defending their own bowers. By choosing a male with a well-ornamented bower, the female is therefore choosing a male with proven fighting ability. Another example of an 'honest' signal is the association between body size and pitch of calls in amphibia (Davies and Halliday, 1978). Small males simply cannot produce low frequency sounds so the pitch of a call is an honest indicator of body size.

There are, therefore, two views about the evolution of signals – the Dawkins–Krebs view that stresses the manipulatory nature of signals, and the Zahavian view that signals are essentially honest.

Which one is correct? As I stressed before, I do not think these two views are incompatible: both could be correct. Dawkins and Krebs discussed a coevolutionary process without specifying an end-point, whereas Zahavi was concerned mainly with the end-point itself, so it is possible to imagine an evolutionary arms race of manipulation and sales resistance which ends up with honest signalling. This cannot be the whole answer, however, as the cuckoo example illustrates. Cuckoos are not honest signallers.

It is, incidently, interesting to note that when Zahavi first proposed this view of signals being honest through what he called the 'handicap principle' (1975), he was met with ridicule. Now, various theoretical models have been produced (e.g. Kodric-Brown and Brown, 1984; Enquist *et al.*, 1985; Grafen, 1990) which show that Zahavi's ideas are quite plausible. By postulating a graded handicap (related to underlying quality) instead of the original 'fixed' handicap (handicap the same regardless of quality of the individual), the idea of honest signalling has become widely accepted.

I want now to turn to a second aspect of the modern view of signals – that of the function of signals. Here the views that Tinbergen held have been largely superseded. In particular, a variety of game theory models pioneered by Maynard-Smith (1982) and Maynard-Smith and Parker (1976), have radically changed the way we see courtship and threat signals. For example, whereas Tinbergen saw courtship signals as serving to arouse members of the opposite sex, to synchronize mating activities and to maintain the pair bond, we would now say that the signals allow members of the two sexes to assess each other as regards paternity, fidelity and quality. Similarly, whereas Tinbergen stressed the universal importance of accurate transmission of information about intentions, it is now clear that many so-called threat displays are concerned with the assessment of fighting ability (RHP or resource holding potential – Maynard-Smith and Parker, 1976).

Two examples can be used to illustrate the role of signals in assessment. Hamilton and Zuk (1982) proposed that females may use the elaborate plumage or other ornaments of males to assess the extent to which males are resistant to disease. Zuk *et al.* (1990a, b) showed that female red jungle fowl choose males with large red combs, brightly coloured eyes and bright feathers on their hackles and flanks. If male jungle fowl are infected with a nematode, *Aspiridia galli*, it is those same features that are most affected by the parasitic infection, whereas other features of the body are much less affected. So it appears that the females are using as courtship signals those features of a male that most reliably indicate his resistance to disease.

In the field of threat signals, the roaring rate of red deer appears to be used by other males to assess fighting ability (Clutton-Brock and Albon, 1979). The ability to roar at a high rate and the ability to fight are closely linked because they are both exhausting and both use the same muscles. A stag in poor condition is simply unable to do either, so ability to maintain a high roaring rate is an honest indicator of fighting ability.

Another area in which game theory models have contributed is in the debate about the extent to which signals convey information about intentions. Tinbergen, or course, saw this as an important feature of communication. Maynard-Smith (1982), however, initially argued the opposite. He concluded that the evolutionarily stable strategy in a fight would be never to give away intentions at the beginning of an encounter, but to play 'poker faced'. Caryl (1979) supported this by citing studies by Stokes (1962), Dunham (1966) and Andersson (1976) which showed that threat displays were not very accurate predictors of an actor's intentions. However this did not seem to be a universal finding. Whereas Jakobssen *et al.* (1979) and Simpson (1968) studied examples in which fish do not communicate their intentions during a fight, Turner and Huntingford (1986) found another case where they do. Theory clearly did not always fit the data. Then Enquist and Leimar (1983) produced another game theory model in which (using the Zahavian idea of a cost to signalling) they showed that it can pay participants to communicate their intentions about a fight. Both empiricists and theoreticians could relax. Whether animals played poker faced or communicated their intentions, there was a theory on hand to account for all the data. We need much more detailed analysis of the cost of signals in relation to their benefit before we can predict accurately what animals will do in any given circumstances.

When we turn to modern ideas about the variety and design of signals, we find a shift in emphasis from Tinbergen's account in terms of causal explanations to recent interpretations in terms of function. For example, Morton (1975) and Nottebohm (1975) have looked at the nature of songs in different species and populations of birds in relation to the acoustic properties of the environment in they were living. My student, Mac Hunter, and I (Hunter and Krebs, 1979) recorded the songs of great tits living in two contrasting habitats, dense forest and open parkland, in different parts of their range. We thus had recordings of a single species in Spain, Iran, Greece, Oxfordshire, Sweden, Norway and Poland. We found a consistent pattern: birds living in open habitats tended to have a higher maximum frequency, a greater frequency range and greater complexity than birds living in forest habitats. In other words, we had found a wide ranging ecocorrelate for the design of signals. Quite why it has arisen, is not clear: it may be to do with the attenuation of different sounds in different areas (Morton, 1975) or with degradation of sound through reverberation (Wiley and Richards, 1978), but the signals are clearly 'designed' differently in different habitats.

As well as looking to the environment to explain signal design, we can also turn to what can be called 'receiver psychology'. Guilford and Dawkins (1991) have argued that in order to understand why signals are the way they are, we have to know about the psychology of the animals that are responding to the signal. What reactors find easy to detect, discriminate or remember about a signal will be an important selective agent in the evolution of signal design. A recent study by Ryan *et al.* (1990) makes this point simply but effectively. They studied two species of Central American frog, *Physalaemus pustulosis* and *P. coloradorum.* In *P. pustulosis,* the male has a courtship call, called the 'chuck' call, which is crucial for getting the female to mate. The chuck call has its main energy in the range 2–2.5kHz, and neatly matching this, the female's auditory system is tuned to about the same frequency - her basilar papilla is particularly sensitive to this same frequency range. Even more interesting was that in the closely related species, *P. coloradorum,* the male does not give the chuck call, but the female does have the same frequency sensitivity. Ryan *et al.* argue that the sensory bias in the female preceded the evolution of the call in the male, while this is not the only interpretation of this result (Pomiankowski and Guilford, 1990), if it is correct it would provide a very neat example of how the design of the signal has evolved around receiver psychology.

In Zahavi's view, the design of signals would, as we have already seen, depend on how the signals honestly indicate differences in quality. Not only would he explain stotting as an honest indicator of running speed, he would also explain the design of all other signals in terms of indicators of quality. It might be quite hard to explain, say, the difference in song organization between a blackbird and a song thrush, in terms of honest indicators of something that was ecologically important and different between those two species. Nevertheless, the idea that signals do honestly indicate some quality is an intriguing and heuristically powerful one.

In summary, the Tinbergen–Lorenz view of communication and the newer view contrast in several different ways. Apart from the few examples referred to by Huntingford (Chapter 4), the study of the causation of signals in the sense that Tinbergen meant, has virtually stopped. There is also now much more emphasis in coevolutionary processes. Although it is possible to infer from some of Tinbergen's writings that he did understand that coevolution occurred in signal evolution, it is not very explicit. He wrote 'All these changes seem to aim at one end, adaptation to the responsive capacities of the reactor' (these changes being the changes of ritualization) (1952). He may have understood something of the evolutionary interactions between signaller and receiver but it was not a major part of his explanation of why signals became ritualized. Nor did he place any emphasis on the distinction between the different coevolutionary routes taken by cooperative and noncooperative signals, that Dawkins and I have stressed (1984). The idea of signalling and honesty is also completely absent; so is the link between ecology and the design of signals. Ironically, although Tinbergen was usually so clear about the distinction between the four questions (Tinbergen, 1963), he confused functional and causal explanations when it came to courtship displays, for example, he used basically mechanistic accounts of maintaining the pair bond or leading to sexual arousal. Perhaps we are now more careful to make the distinction. Rather than being satisfied with an account of courtship as serving to maintain the pair bond, we ask why, in evolutionary terms, this is necessary. Many of our present functional explanations of signals in terms of assessment owe a great deal to Maynard-Smith (1982).

Having so far looked back, let us now look forward. There is room for a great deal of improvement in our basic knowledge and our conceptual framework. 'Receiver psychology' is one area that would seem particularly rewarding, as we know surprisingly little about how animals perceive signals. Most of our interpretations are anthropomorphic. The claim that great tit songs are more complex in open habitats than in forest habitats (see above) is based on the fact that it appears that way to me. But how do I know that great tits or chickens or other birds perceive signals in the same way that we do? This is a very important area for future research. Geoff Cynx and my former research student, Danny Weary, (1990; Nottebohm *et al.*, in press) have already shown one way in which it can be fruitfully pursued. They have asked how birds perceive their songs by using operant procedures. For example, in Weary's study (1990) a bird is first trained to respond to a song by making some response that is rewarded. Then songs differing in various ways are introduced and the bird is effectively asked whether it categorizes these sounds as like or unlike the original song.

Another approach can be illustrated by an experiment carried out by Falls, Dickinson and Krebs (1990). We were working on the eastern meadowlark, a species where individual birds may have hundreds of songs in their repertoire. One bird can have in its repertoire some songs that sound very different (to us) and others that seem very

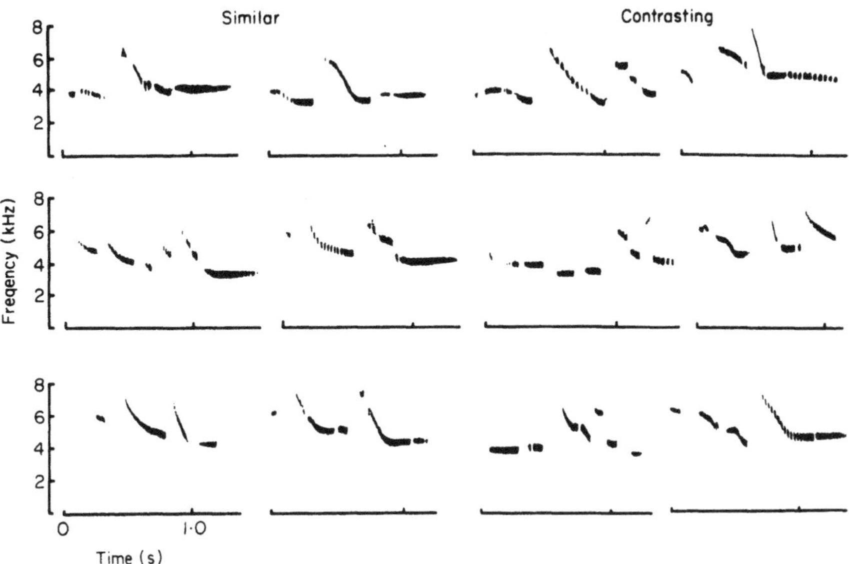

Figure 5.2 Sonagrams of the song types used in constructing the playback experiments on eastern meadowlarks. Each row shows songs from one individual bird's repertoire, classified as to whether they were similar (to humans) or contrasting. (From Falls *et al.*, 1990).

similar (Fig. 5.2) We wanted to know whether the meadowlarks also perceive the songs as either different or similar, and attempted to find out by playing each bird a double playback tape and recording its response. The tape might start off with song A and then switch halfway through to song B. Sometimes A and B were very similar (to us) and at other times very different. What we were looking for was evidence that the birds could detect the changeover from A to B, which would be evidence that they could detect the difference. The answer we got was very surprising. If two songs were taken from the bird's own repertoire, then it was perfectly well able to detect the switch between two 'similar' songs. In fact, for its own songs, there was as much change in behaviour between two songs we had classified as similar as between those songs we had classified as different. However, if the test songs came from the repertoire of another bird that was completely unfamiliar to the test bird, it appeared oblivious to the subtle changes and only responded to the larger changes that were more obvious to us. This study shows us that the perception side of animal communication may be quite subtle, and how, in this case, experience can influence what the animal perceives as being similar or different. It also appears, as suggested by Nottebohm *et al.* (1990) that there is a close link between perception and production.

A final area of animal communication which I see as a challenge for the future comes from Zahavi's (1975, 1987) idea that signal design can be understood in terms of honest signalling maintained by cost. I gave a few examples earlier of how 'honesty' does seem to be a feature of various signals. If this is a widespread phenomenon, the implications for understanding a variety of as yet unexplained signals are immense.

ACKNOWLEDGEMENTS

I thank the Royal Society for financial support. Marian Dawkins turned a recording of my lecture into a manuscript with remarkable skill and energy: I am most grateful to her.

REFERENCES

Andersson, M. (1976) Social behaviour and communication in the Great Skua. *Behaviour*, **58**, 40–77.

Borgia, G. (1985) Bower quality, number of decorations and mating success of male satin bowerbirds (*Ptilinorynchus violaceus*): an experimental analysis. *Animal Behaviour*, **33**, 266–71.

Caryl, P.G. (1979) Communication by agonistic displays: what can games theory contribute to ethology? *Behaviour*, **68**, 136–69.

Clutton-Brock, T.H. and Albon, S.D. (1979) The roaring of red deer and the evolution of honest advertisement. *Behaviour*, **69**, 145–70.

Cullen, J.M. (1966) Reduction of ambiguity through ritualisation. *Philosophical Transactions of the Royal Society B* **251**, 363–74.

Davies, N.B. and Halliday, T.R. (1978) Deep croaks and fighting assessment in toads, *Bufo bufo*. *Nature, Lond.* **274**, 683–85.

Dawkins, R. (1976) *The Selfish Gene*, Oxford University Press, Oxford.

Dawkins, R. and Krebs, J.R. (1978) Animal signals: information of manipulation?, in *Behavioural ecology: an evolutionary approach* (eds. J.R. Krebs and N.B. Davies), Blackwell Scientific Publications, Oxford, pp. 282–309.

Dunham, D.W. (1966) Agonistic behaviour in captive rose-breasted grosbeaks, *Pheucticus ludovicianus* (L.) *Behaviour*, **27**, 160–73.

Enquist, M., Plane, E. and Röed, J. (1985) Aggressive communication in fulmars (*Fulmarus glacialis*) competing for food. *Animal Behaviour*, **33**, 1007–20.

Enquist, M. and Leimar, O. (1983) Evolution of fighting behaviour: decision rules and assessment of relative strength. *Journal of Theoretical Biology*, **102**, 387–410.

Falls, J.B., Dickenson, T.E. and Krebs, J.R. (1990) Contrast between successive songs affects the response of eastern meadowlarks to playback. *Animal Behaviour*, **39**, 717–28.

Fitzgibbon, C.D. and Fanshawe, J.H. (1988) Stotting in Thomson's gazelles: an honest signal of condition. *Behavioural Ecology and Sociobiology*, **23**, 69–74.

Grafen, A. (1990) Biological signals as handicaps. *Journal of Theoretical Biology*, **144**, 517–46.

Guilford, T. and Dawkins, M. (1991) Receiver psychology and the evolution of animal signals. *Animal Behaviour* (in press).

Hamilton, W.D. (1964) The genetical evolution of social behaviour. *Journal of Theoretical Biology*, **7**, 1–52.

Hamilton, W.D. and Zuk, M. (1982) Heritable true fitness and bright birds: a role for parasites? *Science*, NY. **218**, 384–87.

Hunter, M.L. and Krebs, J.R. (1979) Geographical variation in the song of the great tit (*Parus major*) in relation to ecological factors. *Journal of Animal Ecology*, **48**, 759–86.

Huxley, J. (1966) (organizer) A discussion on ritualization of behaviour in animals and man. *Philosophical Transactions of the Royal Society*, B. **251**, 247–526.

Jakobssen, S., Radesater, T. and Jarvi, T. (1979) On the fighting behaviour of *Nannacara anomala* (Pisces, Cichlidae) *Zeitschrift für Tierpsychologie*, **49**, 210–20.

Kodric-Brown, A. and Brown, J.H. (1984) Truth in advertising: the kinds of traits favoured by sexual selection. *The American Naturalist*, **124**, 309–23.

Krebs, J.R. and Dawkins, R. (1984) Animal signals: mind reading and manipulation, in *Behavioural Ecology. An evolutionary approach*, 2nd. edn, (eds J.R. Krebs and N.B. Davies). Blackwell Scientific Publications, Oxford, pp. 380–402.

Lloyd, J.E. (1979) Mating behaviour and natural selection. *Florida Entomol*, **62**, 17–34.

Maynard-Smith, J. (1982) *Evolution and the theory of games*, Cambridge University Press, Cambridge.

Maynard-Smith, J. and Parker, G.A. (1976) The logic of asymmetrical contests. *Animal Behaviour*, **24**, 159–75.

Morton, E.S. (1975) Ecological sources of selection on avrian sounds. *The American Naturalist*, **109**, 17–34.

Nottebohm, F. (1975) Continental patterns of song variability in *Zonotrichia capensis*: some possible ecological correlates. *The American Naturalist*, **109**, 605–24.

Nottebohm, F., Alvarez-Buylla, A., Cynx, J. *et al.* (1990) Song-learning in birds: the relation between perception and production. *Philosophical Transactions of the Royal Society* B **329**.

Ryan, M.J., Fox, J.H., Wilczynski, W. and Rand, A.S. (1990) Sexual selection for sensory exploitation in the frog *Physalaemus pustulosis*. *Nature*, Lond., **343**, 66–7.

Simpson, M.J.A. (1968) The display of Siamese fighting fish, *Betta splendens*. *Animal Behaviour Monographs*, **1**, 1–73.

Stokes, A.W. (1962) Agonistic behaviour among blue tits at a winter feeding station. *Behaviour*, **19**, 208–18.

Tinbergen, N. (1952) 'Derived' activities, their causation, biological significance, origin and emancipation during evolution. *Quarterly Review of Biology*, **27**, 1–32.

Tinbergen, N. (1963) On the aims and methods of ethology. *Zeitschrift für Tierpsychologie*, **20**, 410–33.

Turner, A. and Huntingford, F.A. (1986) A problem for game theory analysis: assessment and intention in male mouthbrooder contests. *Animal Behaviour* **34**, 961–70

Weary, D. (1990) Categorization of song notes in great tits: which acoustic features are used and why? *Animal Behaviour*, **39**, 450–57.

Wiley, R.H. and Richards, D.G. (1978) Physical constraints in acoustic communication in the atmosphere: implications for the evolution of animal vocalizations. *Behavioural Ecological and Sociobiology*, **3**, 69–94.

Zahavi, A. (1975) Mate selection – a selection for a handicap. *Journal of Theoretical Biology*, **53**, 205–14.

Zahavi, A. (1979) Ritualization and the evolution of movement signals. *Behaviour*, **72**, 77–81.

Zahavi, A. (1987) The theory of signal selection and some of its implications, in *International Symposium of Biological Evolution*, (ed. V.P. Delfino), Adriatica Editrice, Bari.

Zuk, M., Johnson, K., Thornhill, R. and Ligon, D.J. (1990a) Mechanisms of female choice in red jungle fowl. *Evolution*, **44**, 477–85.

Zuk, M., Thornhill, R., Ligon, J.D. *et al.* (1990b) The role of male ornaments and courtship behaviour in female mate choice of red jungle fowl. *The American Naturalist*, **136**, 459–73.

6
The nature of culture

JUAN D. DELIUS

'Townley . . . had said one word only, and that one of the shortest in the language, but Ernest was in a fit state for innoculation, and the minute particle of virus set about working immediately'. (Samuel Butler, 'The Way of All Flesh', 1903).

CULTURE AND BIOEVOLUTION

Niko Tinbergen laid great stress on the essential importance of cultural evolution for the understanding of human behaviour although he never, of course, made it a central subject of his research interests. In a short cautionary note about the future of humanity he wrote for example that 'our unique position in the modern world is due to the consequences of our cultural evolution, which . . . has . . . progressively . . . superimposed (itself) on our still ongoing genetic evolution' and that 'we transfer . . . , from one generation to the next, not only our genetic heritage but also (our) accumulated non-genetically acquired . . . experience' (Tinbergen, 1977, see also Tinbergen, 1976). Niko's insights into the details of the processes of cultural evolution went much further than his writings reflect, however. A casual but memorable conversation between him and Konrad Lorenz in Stuttgart, Germany in 1959, at which I happened to be present, revealed that clearly. The role of song behaviour as a species-isolating mechanism in some sympatric birds had somehow cropped up. They were considering the selective forces that might have shaped the divergence of song patterns in such situations when Niko raised the important question: Selection of what? Surely not genes since the song of these birds was likely to be learned, not innate. Konrad suggested that song traditions were being selected but Niko considered that it might be better to think of song memory traces being selected. Would memory evolution always cooperate with gene evolution? In the 20 minutes or so that followed, without ever mentioning the word culture, if I remember correctly, they had worked out between them the essentials of a modern theory of cultural evolution. There were however several later dialogues at the Ravenglass field camp where Niko actually pursued similar arguments explicitly in relation to human culture. In recent years, I have tried again and again to recapture some of the threads of those discussions. Needless to say, besides imperfect recollections, this essay also takes into account some of the theoretical ideas and empirical findings (these latter are still scarce) of many scholars that since Tinbergen's times have made cultural evolution a more definite subject of their interest (notably Campbell, 1969; Bajema, 1972; Dawkins, 1976; Cavalli-Sforza and Feldman, 1981; Lumsden and Wilson, 1981; Boyd and Richerson, 1985).

It is time to heed the maestro's ever-recurring admonitions about defining one's terms. Indeed, the word culture is commonly used with several meanings. For the purposes of this essay it is necessary to circumscribe its definition to behavioural culture. Culture will thus mean here the ensemble of traditional behaviours that is characteristic of a population. Traditional behaviours are those that individuals take over from others through some form of social learning. Sometimes media (for example newspapers, books, television) intervene in this process of transmission. Behaviour patterns that organisms acquire via genetic inheritance (for example hatching, crowing, smiling, crying) or by individual learning (key-pecking, soft landing, nose-picking, masturbating) are excluded by this definition. This trichotomous classification of behaviour is avowedly simplistic but here it is conceptually useful. Some illustrative examples of cultural behaviour could be birds nesting in a traditional area, singing a certain dialect and mobbing particular predators, or humans wearing a particular dress, speaking a certain language, reading a certain bestseller, and worshipping a particular god. Material objects by the above definition are not really part of culture, but they are often convenient referents for the cultural behaviour that produced them or is elicited by them (the book or the clothing in the above examples). The definition however appears to be able to accommodate without much strain most other, less tangible 'contents' of culture commonly listed in anthropology textbooks (Harris, 1987): knowledge, beliefs, rituals, institutions, customs, fashions, symbols, etc.

Culture is clearly not a universal attribute of all organisms. In fact, only two or three decades ago it was thought that only humans were endowed with culture. This opinion is no longer held. Cultures, or at least protocultures, have now been documented in many animals (Bonner, 1980), but it is also true that the phenomenon only occurs in a proportion of the more advanced species. The permissive trait, as suggested by the above definition of culture, is that they must be capable of social learning, a competence that comes about through biological evolution. Even when this basic capability is already present, culture is the product of a lengthy historical development. The cultural behaviours proper for knights, Scots, or 'yuppies' are clearly not god-given but have developed gradually over a long time. Their common ancestors 50 000 years ago, the Cromagnons, certainly did not show the cultural traits we now associate with these groups. The cultural evolution process has long been recognized as having at least superficial similarities with that which drives biological evolution. A more thorough anlaysis of the potential analogy has however begun only recently. The intention of this essay is essentially to explore how far the similarities between biological and cultural evolution actually go.

Consideration of how the capacity for culture might have emerged affords the opportunity to briefly recapitulate the salient characteristics of biological evolution. It is a game that genes are simply fated to play as a consequence of their particular molecular properties (Dawkins, 1976). The essential property is that they are capable of self-replication that is not always perfect. Since the gene mutants that arise in this way interact with and compete for an environmental niche in which to survive and replicate, it follows that they will frequently differ in replication potential, that is in fitness. The consequence is gene selection. In different niches different mutations may be the fittest, and this eventually results in speciation. Genes capable of instructing the synthesis of buffering devices, a membrane or even a soma, against environmental variability are likely to have fitness advantages. Gene mutants that could instruct devices that added motility and also sensitivity, behaviour in short, would, given the right circumstances, be even fitter. Which responses followed which stimuli was initially determined exclusively by genetic instruction (innate behaviour). In environments that were more variable over time and space, selection pressure arose for mutations that could instruct neural structures capable of learning. The capacity, for example, to attach existing responses to arbitrary stimuli that happen to be predictive of fitness-influencing events (classical conditioning) or to modify behaviour in such a way as to influence the likelihood of such events (instrumental learning) as a consequence of individual experience obviously magnifies the adaptability of organisms (Staddon, 1983). Mutants extending memory capacities so that a representation of the environment and the self was feasible signified a further bonus. An internal off-line behaviour simulation would become possible, which could even include creative innovation (insight learning). Generally, learning is a device instructed by genes that allows the individual to acquire knowledge about the world and itself over and above that implicitly contained in its genetic code. It is naturally also associated with costs, such as more complex brains and some behavioural instability.

SOCIAL LEARNING AND CULTURE

When an organism acquires behaviour through individual learning, the process is often lengthy, risky and laborious. Genes giving rise to structures that enabled animals to take over the already extant experience of conspecifics would often yield a fitness advantage. The essential characteristic of social learning (imitation, observation, instruction learning) is indeed that individuals in one way or another take over knowledge from others. The precise mechanisms supporting this transmission of information vary considerably (Zentall and Galef, 1988).

Pigeons, as do many other species, tend to breed at sites close to those where they themselves were bred. Successive generations keep to traditional breeding grounds and this is not for want of mobility, as they may migrate far in between breeding. Through an imprinting-like process, juveniles store information about the location where they grow up. This enables them to navigate back to the home area later (Schmidt-Koenig, 1965). It is not only the geographical location to which the young birds imprint but also to the particular habitat in which they were raised, to cliffs or buildings for example. As adults they will then show a preference for the same type of surroundings in which they grew up (Delius, unpublished observations; Klopfer and Hailman, 1965). This does not come about by the youngsters directly imitating the parents but rather by the parents bringing their youngsters up where they can only learn about one thing. There is often a debate about whether this represents true social learning, but it certainly serves to maintain familial traditions.

Seeing several flock members fly to a particular site usually induces other pigeons to follow. Such socially facilitated behaviour need not involve any learning, often being based on innate behaviour (Tinbergen, 1953). But at the same time follower pigeons can hardly fail to learn about the association between, for example, granaries and grain, something that the leading pigeons already knew (Murton *et al.*, 1972). In some altricial species the parents lead their young to sites where the food that they themselves prefer predominates and each of the young learns on its own to find and deal efficiently with these items (Subowski, 1989). Thus again traditions may simply be maintained by parents biasing the learning opportunities of their offspring. Oystercatchers prey on mussels using one of two techniques, stabbing or hammering. Youngsters appear to learn the particular technique that the parents used through observation and participation (Norton Griffiths, 1967; but see Meire and Ervynick, 1986). In a similar context female cats with kittens may even display behaviour analogous to teaching by repeatedly bringing home and releasing live prey just to catch it again. Kittens gradually join the repeated chases and eventually learn to to do the final killing themselves (Chesler, 1969; Ewer, 1969).

Contrived non-natural modes of food gathering have been experimentally arranged to arise through imitative instrumental learning in several species. Having seen another pigeon obtain food from an electromagnetic dispenser after performing the somewhat arbitrary behaviour of pecking an illuminated disc considerably facilitates the subsequent acquisition of that same skill by observer pigeons. A number of experiments show that the information the observer acquires can be multifarious: knowledge of the fact that food is to be found in the particular environment; that it is available at a particular place within that environment; that performance of certain acts make that food more available, and so on. It is rare, however, for an observer to produce the correct food-yielding behaviour on the first attempt. Rather, the observer is only quicker at learning what has to be done. An opportunity to perform the target behaviour while observing is helpful but learning is also facilitated when key-pecking is only possible after a delay (Alderks, 1986; Biederman *et al.*, 1986; Hogan, 1986).

Lefebvre (1986) set up an artificial feeding tradition among a flock of urban pigeons. They captured a few flock members and trained them to feed by piercing a tight paper sheet covering their food troughs. When they were released and were back with their flock, the latter was offered paper-covered troughs. The trained birds immediately began to pierce and feed. Soon no fewer than two dozen other birds had acquired the paper-piercing technique. In a control flock that did not have pretrained demonstrators it took almost three times as long before a bird 'invented' paper piercing by himself, but once that happened, the cultural trait spread just as fast among that flock. Novel feeding cultures occasionally arise naturally among free-ranging animals. A Japanese female macaque named Imo discovered in 1953 that sweet potatoes which she had accidently dropped in a brook tasted better than unwashed, earthy ones and began to actively wash them before consumption. By 1958 all the younger monkeys in her band had imitated her: a potato-washing culture had arisen. Other Japanese monkey groups have never developed this tradition in spite of similar opportunities, but some have developed other local traditions (Nishida, 1987).

Social learning can also be mediated by classic conditioning. Wild-caught blackbirds exhibit mobbing behaviour when they see an owl. Curio *et al.* (1978) arranged it that a predator-naive, hand-raised blackbird saw a novel inoffensive plastic bottle, while an experienced bird actually mobbed an owl. Mobbing acts as an unconditioned stimulus: the naive blackbird began to mob too (unconditioned response, *see* social facilitation above). The bottle functioned as a conditioned stimulus as it always preceded and accompanied the owl-mobbing by the model. After a few pairings, conditioned mobbing could be demonstrated. In the absence of any model the observer blackbird now mobbed whenever the bottle was presented. The acquired bottle-mobbing habit could in turn serve as model for new observers. Thus a novel mobbing tradition or culture had been set up among blackbirds.

The cultural nature of songbird song is so well known that only the most essential characteristics and the most common variants of the underlying process will be mentioned. Song varieties that young birds hear from their father or his neighbours are memorized during a critical period. This imprinting usually takes place before the youngsters can themselves sing. Later the sub-adults learn to match the auditory template with their own song. Barring occasional errors, this leads to a fairly accurate replication of the songs originally heard. Normally this mode of song acquisition leads to the emergence of song dialects characteristic of local populations within a given species (Barker and Cunningham, 1985; Catchpole, 1986). Some species continue to be able to acquire songs all their lives and some may even imitate the sound of other species. There are almost 4000 species of songbirds and it is resonably certain that, except for a few, all sing according to traditions. This suggests that their common ancestor, the original songbird living about 40 million years ago, already had a song culture, long before our primate ancestors had any culture.

Most other taxonomic bird groups have innate songs but parrots are also well known for learning vocalizations by imitation. Since parrots are only distantly related to songbirds, but closely related to birds that have innate vocalizations, their cultures must have arisen, independently (Kroodsma and Miller, 1982).

In the human species, social learning is almost obligatory and often involves yet another, more refined variant. Having previously seen somebody light a camp fire is certainly of assistance if one had to do the same thing for the first time. The demonstrator performs on the basis of information stored in his or her memory. The behaviour thus produced is seen by the observer, who in turn stores these perceptions in memory, only to convert them again into behaviour at a later date. In such cases, the model has to perform for the observer to be able to memorize the actions. Language, however, enables an energetically more economical form of memory transfer. Simply being told how to light a camp fire is usually sufficient for a reasonable emulation. In most situations where human individuals adopt behavioural traits from others, language plays an important supportive, if not sole mediating role.

It has, indeed, been argued that language might have arisen evolutionarily as an extension of social learning, as a vehicle for instruction, whereby recipes for behaviour could be transmitted in an abstract symbolic way, in a code that might be related to that of memory (Catania, 1985; Delius, 1990). For a *Homo habilis* hunter a million years ago it must have been awkward to demonstrate to novices how to stalk antelopes; it would have been much easier to tell them how it should be done. Linguistic messages function like an almost effortless short circuit between the memories of individuals. Accordingly, it accounts for much of the sophistication of human culture *vis à vis* animal culture. Indeed, linguistic communication, spoken or written, is sometimes the only medium by which many human traditions can be transmitted. Writing greatly amplifies the multiplicative power of language. Most importantly, it disposes of the necessity for the model and the observer to have to coincide in time and space for transmission to be possible. In a way, it makes social learning possible in an asocial setting. Modern communication media extend this even further.

MNEMOBIOLOGY AND MEMES

Generally, learning, whether individual or social, can be conceived as a process whereby experiental information is stored into memory. Cultural traits, defined earlier as behavioural items acquired through social learning, are therefore also represented as particular contents in the memory of the individual bearers of culture.

Information storage is necessarily dependent on physicochemical state changes in memory-supporting structures. According to neurobiological findings, learning (social or otherwise) leads first to volatile changes, lasting only tens of seconds or minutes. Only when learning is sufficiently incisive in one way or another, will the relevant memory traces be consolidated into a more durable format lasting months or years. Culture as a rather persistent phenomenon is obviously heavily dependent on these long-term memories.

Long-term memory, according to current evidence, is laid down as structural brain modifications. Memory deposition chiefly involves changes at the level of the interneuronal transmission sites, the synapses (Morris *et al.*, 1988; Dudai, 1989). Due to the particular patterns of coactivation of pre- and postsynaptic neurons arising during learning, some of these synapses pass from a state of relative inefficiency to a state of relative efficiency, from an inactivated to an activated condition, rather like bits in computer memory that are set from a 0 off-state to a 1 on-state. In some instances, learning even seems to lead to the budding of additional synapses and to the growth of neuronal ramifications supporting them (Horn, 1986). However, much as computer memory also stores information when bits pass from a 1 to a 0 state, there seems to be instances where learning is associated with synapses passing from an activated to an inactivated state or even disappearing, sometimes together with their supportive structures (Wallhäusser and Scheich, 1987). That specifically social learning is also associated with such neural modifications is best documented for songbird vocalizations (Devoogd *et al.*, 1985).

A pigeon brain contains perhaps 10^{10}, the human brain maybe 10^{15}, synapses that are variable in the above sense. Such plastic synapses have to be thought of as the critical components of neural networks functioning as associative arrays. It has been shown mathematically and confirmed empirically that neural networks incorporating large numbers of modifiable junctions are able to store vast quantities of information in a very organized manner. An important property of associative network storage is that the information is content- and not address-retrievable (as in computers), and furthermore that it is stored in a highly distributed but still partially overlapping way (Palm, 1982). Special versions of these networks show interesting additional properties, such as being capable of self organization, stimulus pattern categorization, pattern completion or complex stimulus-response conversions (Kohonen, 1984; Rummelhardt and McClelland, 1986).

Any cultural trait taken over by a given individual from another individual must accordingly be thought of as the transfer of a particular pattern of activated/inactivated synapses from the associative networks of one brain to another. Different traits must be thought of as being coded by topologically different synaptic patterns, that is, a given cultural trait borne by an individual is encoded informationally as a particular configuration of modified synapses in his or her brain (Figure 6.1). Naturally the synaptic constellation that a trait has in one brain will not be geometrically arranged in exactly the same way as the pattern that the same trait has in another brain: the brains of different individuals are likely to be too different for that. Functionally however, the two patterns could still be

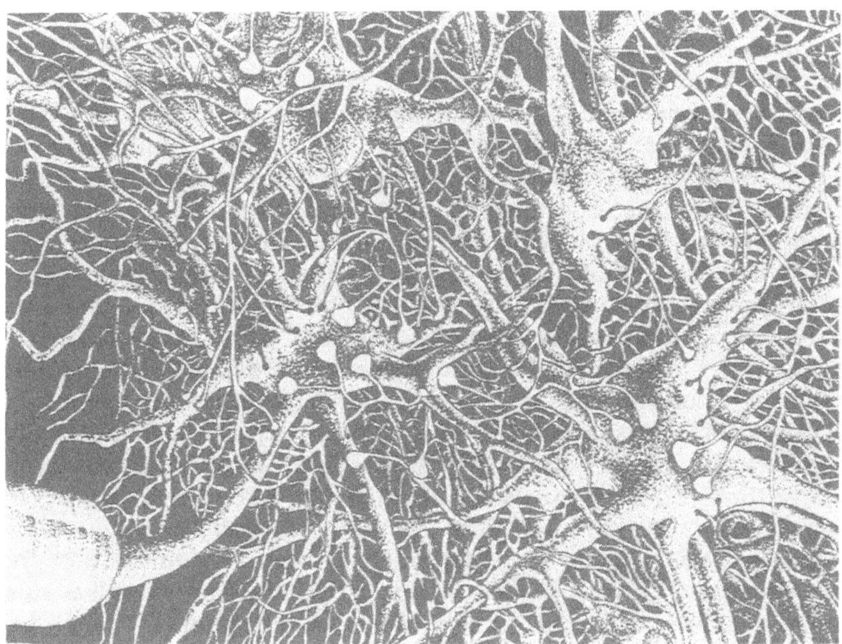

Figure 6.1 A meme as a constellation of activated neuronal synapses lodged somewhere in the brain of an individual

equivalent when effectively identical traits were represented in memory. In any case, following Dawkins (1976), synaptic patterns that code cultural traits will be called memes, by analogy with the molecular patterns that code biological traits and which are called genes.

The process of cultural heritage can be seen as passing on these synaptic constellations or memes from one individual to another, or, and that is important, to several other individuals. Obviously it is not a bodily replication of material structures, as is usual with genes, but social learning nonetheless brings about a multiplicative transfer of equivalent structures. Even among genes, however, replication can sometimes be less than direct. Retrovirus RNA genes for example depend heavily on mediation by host cellular machinery for reproduction (Davies *et al.*, 1980). Memes are capable of instructing, not protein synthesis as genes do, but behaviour. However, genes can do that too indirectly through protein synthesis. On the other hand meme replication, by involving neurostructural modifications, is invariably associated with the induction of protein synthesis.

Genes can be conceived as specific molecules that code information according to a well understood scheme. Until the 1950s however, genes were largely hypothetical constructs that could not be linked to anything more specific than approximate sites on chromosomes. Memetics is not yet as advanced as genetics. Memes are still largely abstract inferential entities, though we know that they are information coded in neural structures. It is possible and even likely that the memetic code is less universal and more complex than the genetic one. However, the way in which genes code innate behaviour, say the suckling reflex or the crying response of babies, is anything but simple and unitary. Some authors (Cavalli-Sforza and Feldman, 1981; Boyd and Richerson, 1985) nevertheless prefer to speak of cultural traits or cultural variants rather than of memes (or culturgenes; Lumsden and Wilson, 1981). That seems linguistically cumbersome. Memes stress the transmittance of coded information rather than of behaviour itself, which is a physical impossibility. In genetics it is conceptually important to separate the phenotypic characters and the genes that determine it. It must be admitted, though, that because of the unsatisfactory state of memetics we are still as rule forced to allude to memes by way of trait descriptions.

The important point is that memes have the same essential properties that make genes the key protagonists of an evolutionary process (Dawkins, 1976). They are obviously capable of replication, even if in a roundabout way. Replication is reasonably faithful but not perfect, that is, memes mutate. New song variants arise among songbirds, new words are coined, new rituals are derived and new fads emerge among humans. Not all meme variants are equally effective in reproducing themselves. Some memes spread rapidly in a population, others become extinct. 'I know something, but I won't tell anybody' is a nonstarter as memes go, but 'I will tell you a sure way to save tax' stands for a meme likely to multiply (Ball, 1984). Different memes have different cultural fitnesses, that is, memes vary in their potential to produce memetic offspring.

GENE/MEME SYMBIOSES

There are two characteristics that seem to make the memes of an individual plainly different from his genes. One is that his memes are lodged in the brain that was instructed by his genes, i.e. the former depend on the latter. The other is that the individual's genes come from his parents, but his memes can and often do originate from other conspecifics besides his parents (friends and teachers for example). A host–guest relationship is not at all exceptional among genes themselves, the genes of many symbiotic organisms residing in organisms instructed by the genes of other organisms. Any average human, for example, is normally host to billions of symbiotic organisms belonging to perhaps a thousand different species (Hohorst, 1981). His phenotype is not determined by his human genes but also by the genes of all the symbionts he happens to be infected with. The symbiont species an individal carries usually have a very varied provenance, with only a few being likely to have come from his parents.

Three kinds of organismic symbioses are distinguished: mutualism, where both partner species benefit from the association in terms of Darwinian fitness (gastric flora of ruminants, for example); commensalism, where the fitness of the guest is furthered by the association with little if any fitness cost or gain to the host (intestinal flora of man, for example); and parasitism, where the symbiont profits on fitness at the expense of the fitness of the host (tapeworms in humans, for example; Smith and Douglas, 1987). A striking instance of a symbiosis of the mutualistic kind are the mitochondria, extranuclear cytoplasmatic organelles of eukaryotic organisms. Mitochondria have their own DNA, which replicates independently of the host-cell DNA. The ancestors of mitochondria some 2 billion years ago were almost certainly parasites in prekaryotic organisms, much as viruses are parasites in present day cells (Margulis, 1981). Nowadays, mitochondria are highly integrated into the higher organisms and are not capable of independent existence. On the other hand, essential processes of host cell respiration are controlled by mitochondria and accordingly they are prominent in metabollically very active neurons. In fact, even the slightest behaviour, the most fleeting thought produced by a eukaryotic animal absolutely presupposes the activity of these obligatory symbionts. Infection is always through the egg cytoplasm; sperm do not carry mitochondria. Mitochondrial DNA undergoes an evolution of its own. Mutant mitochondria emerge occasionally, may replicate and can be favoured or disfavoured by natural selection. The selective agency is the intracellular environment, largely but not exclusively controlled by the host cells' nuclear genes (Cann *et al.*, 1987).

Many parasitic species can influence the behaviour of their hosts in very specific ways (Moore, 1984). The rabies virus, for example, massively invades the host's salivary glands and causes them to secrete profuse quantities of infected saliva, but it also invades the nervous system and influences its functioning so as to inhibit swallowing and increase aggressiveness. The sick animal therefore bites with a large reservoir of infested saliva and is thus very likely to infect its victim, which as a rule is no relation and might even belong to another species (Christie, 1981). But it is apparent that the rabies virus furthers its fitness at the cost of the host's fitness, partly by manipulating the latter's behaviour.

Parasites inevitably generate selection pressure for host mutants that are somehow able to prevent or inhibit infection and/or multiplication of the parasites. Immunological reactions that counteract infections, but also innate behaviour that minimizes chances of infection (pelage grooming, clear water drinking, avoidance of sick conspecifics, for example) promote host fitness. Conversely parasites are under natural selection to increase their ability to infect and replicate. Since they often reproduce at high rates within the host organism, evolutionary adaptation to the immune reactions of the host (or to drug treatment) can come about by selection of resistant mutants within a single host individual. This often mimics Lamarckian evolution because the 'offspring' infection passed onto the next host expresses what appears to be a trait acquired by the 'parent' infect (antibiotic resistance, for instance; Futuyma, 1986).

There is obviously also selective pressure upon host organisms to develop mechanisms for resisting infection or turning parasite organisms into mutualists, or at least into commensals, if at all possible. A too-virulent parasite can potentially kill its host before there has been transmission to further hosts, and thus kill itself and its kin. Restrained virulence may be a better fitness strategy. Parasites may therefore also be under some selection pressure to become mutualists or at least commensals (Anderson and May, 1982) but it competes with opposing pressures that tend to make parasites more infectious. An improved immunological response by a host will generate a selective pressure favouring parasite mutants that can overcome it. This coevolutionary arms-race game is biased by the fact that the usually smaller symbionts have, in general, a much faster generational turnover than the larger hosts. Consequently the relative rates of adaptation and counteradaptation are in favour of parasites rather than hosts.

In so far as memes are material structures (arrays of modified synapses) that reside in host organisms and can multiply independently of them, they can be viewed as analogous to the genes of symbionts. In particular, memes are similar to the genes of those symbionts that invade the brain of their host and influence its functioning in ways that affect the host's behaviour. Memes are replicating coded information packages, which infect some higher animals and manipulate their behaviour. Furthermore their replication, analogously to that of symbiont genes, is not tied to the gene reproduction of host organisms.

When dealing with organismic symbioses, one distinguishes between the genetic fitness of hosts and the genetic fitness of guests even though they are both intimately intertwined. Similarly it is essential to differentiate between the genetic (biological) fitness of bearers of cultural traits and the memetic (cultural) fitness of the cultural traits themselves (Ball, 1984). As is the case in symbionts, specific memes or memomes (meme complexes, in analogy to genomes) could in principle survive and reproduce to the advantage, indifference or detriment of the genetic fitness of their hosts, so that cultural traits could be mutualists, commensals or parasites.

Initially, at the phylogenetically protocultural stage, memes must necessarily have been mutualists simply because the capacity for culture, the genetically determined ability for social learning, can only have spread within a population of organisms if individuals having that competence were biologically fitter than those that did not have it. The genetic fitness advantage of an individual was determined by the actual cultural traits made possible by that capacity, in other words the first few memes at least must have furthered the hosts genes. Social learning among some early hominids, for example, can be imagined to have promoted efficient modes of hunting, efficient ways of tool-making and efficient styles of communication, giving social learners an edge over less educable competitors. Memes at this early stage can be considered as devices by which genes amplified their fitness (Lumsden and Wilson, 1981). They were then close to the slave-like symbionts of genes, much as mitochondria are today. Slaves though, have a well-known bent towards independence. The question is whether genes can manage to keep memes under control past the initial protocultural stage.

Symbionts, as explained earlier, are normally subject to the evolutionary process within their hosts and not just when they are transmitted between hosts. The same can be said of memes. Variability of covert and overt behaviour due to many causes, but not least to neural system noise, continuously generates new potential memes, or meme mutants, within an individual. While still resident in that same individual these mutants are selected by the same innate processes that make learning generally an adaptive process (Staddon, 1983; Gould and Marler, 1987). During overt learning, any otherwise neutral events that circumstantially precede and predict innately appetitive or aversive events, come to be sought or avoided through classic conditioning. Similarly, any arbitrary behavioural response which happens for some reason to generate or produce such reinforcements is correspondingly enhanced or suppressed through instrumental conditioning. During covert learning (insight learning), any event or response identified by behavioural simulation within the memory model of reality as being likely to lead to reward or punishment is retained in, or rejected from, the mind. Eventually, such imagined behaviour is tested when put into practice. Conversely, any responses or strategies that are no longer effective in yielding appetitive or avoiding aversive reinforcement extinguish and are finally forgotten.

In short, any individual learning (including imagining events, ideating, inventing and creating) is a process based on variation and selection, that is analogous to biological evolution (Pringle, 1951; Changeux *et al.*, 1984; Staddon, 1983). Emergent meme mutants in that context can be considered as having replicated well within a host organism, in the sense that the corresponding memory traces have become better consolidated through redundant storage. Unsuccessful meme mutants in the same context are those that cannot establish themselves in the memory of even one host. Incipient meme variants, like symbiotic gene mutants, are thus selected within their hosts, and not only as they spread to new hosts. Effectively every potential meme undergoes a kind of genetically instituted quality control before it has a chance to be passed to other hosts. The commonly held view that cultural evolution is Lamarckian (Medawar, 1976), that it involves the transmission of acquired traits, ignores that fact that the acquisition of those traits is itself the result of a Darwinian intrahost process.

The variation and selection principle however also operates as memes are transmitted from individual to individual by social learning. Imperfect transmission is clearly a frequent source of meme mutation. Potential hosts on the other hand are not completely passive regarding the memes on offer. Host bioevolution can be expected to have ensured that recipient individuals are choosy as to which meme variants they pick up. Young songbirds as a rule only imprint on songs similar to those typical of their species. Obviously an innate mechanism precludes the acceptance of songs that are too aberrant (Catchpole, 1986). A frequent assessment criterion among human seems to be how many carriers of a given meme offer themselves as models. If many conspecifics exhibit a given cultural trait the likelihood that the meme is biologically adaptive is high. If a meme was drastically unadaptive it would literally kill off its hosts, so reducing their number. Another sign of biological meme quality are obvious signs of fitness exhibited by the bearer of a given trait. If an individual is visibly successful, being, for example the strong alpha male of a group of primates, having access to choice food and many females, then the memes he carries are likely to be fitness-promoting and worth acquiring (Boyd and Richerson, 1985).

Even when taken over from other hosts, memes will still have to be consolidated and maintained in the memory of the recipient. Much the same selection mechanisms that operated on the memes generated within an individual, as discussed earlier, will also apply in this situation. There are parallels in symbiont biology. The gene-instructed immune system is an impressively sophisticated mechanism designed to select symbiont variants in the interest of host fitness. It is effective in censoring symbiont varieties, independently of whether they arise through mutations within an individual or whether they enter that individual by infection.

Since genes create the environment on which memes thrive, they should in principle be able to select memes such that only those contributing to their biological fitness can survive and proliferate. Culture at this stage of the game is only one more strategy 'invented' by some genomes to succeed in the harshly competitive game of phylogeny.

PARASITIC MEMES

How discerning can gene-instructed brains be about the memes they are prepared to harbour? The corresponding filter mechanism for symbionts, the immune system of vertebrates, definitely falls short of being perfect. Commensal and even parasitic organisms often get past its scrutiny and manage to infect some, or even many, individuals. Could at least the occasional commensal meme arise and spread in a similar way? Even though penalized by the need for a larger brain, an increased ability for social learning, i.e. an augmented capacity for memes, has obviously conferred a biological advantage in some higher animal species. As long as most of the larger number of memes thus made possible were advantageous to the host, gene selection for indiscriminate reduction of general imitation learning capacities would be weak. Selection for gene mutants that instruct mechanisms capable of controlling more specifically which memes to accept and which to reject would arise, however. The preceding section sketched some mechanisms that could perform this function. But, discrimination between similar meme alleles, some of which promote the genetic fitness of their hosts and others that do not, can demand very clever decision mechanisms. Their instruction is likely to require a slow-to-evolve cooperation of many genes.

If memes were solely cooperative with genes, one would expect the transmission of the former to be closely coupled with the transmission of the latter. According to the genetic selfish altruism principle (Hamilton, 1964), the transmission of mutualist memes should occur mainly between genetically related individuals and less between unrelated individuals. Among primitive cultures most biologically beneficial memes are indeed transmitted from parents to children, and only a few such memes are passed on between non-kin (Hewlett and Cavalli-Sforza, 1986). Unrelated individuals on the other hand are likely to be genetic competitors and should accordingly be suspected of attempting to pass harmful memes to each other. That may be why the search for a solicitor, a stockbroker or a physician among one's relatives is commonplace when one needs critical advice. Still, the fact is that even in protocultures such as those of songbirds, memes are being passed among genetically unrelated individuals.

Mutant memes not contributing to the genetic fitness of their hosts can proliferate if their cultural fitness is high, that is if they are 'catchy', if they can overcome the above-mentioned filters and do not appreciably impair their host's biological fitness. A few biologically useless but inoffensive cultural traits embedded among many cultural traits promoting gene fitness will not generate a strong genetic selective pressure towards mechanisms ensuring their removal. Commensal memes seem an almost inevitable development in any advanced culture. Even in the case of bird culture there are considerable doubts whether all song variants exhibited by a given species are associated with a definite advantage in biological fitness (Barker and Cunningham, 1985).

In human culture biologically innocuous fads or crazes of one kind or another are certainly legion. Pointed rather than round collar tips, two- rather than three-button jackets seem unlikely to make any difference to the survival and the reproduction of wearer, even though generally dressing to keep warm certainly does. Interestingly such commensal memes often occur linked to other memes whose function it is to attach purported biological significance to these. Pointed collars and three-button jackets, for example, were said to be indicative of virility when they were fashionable. On a more global scale, music, literature, and the arts as a whole involve large and complex meme ensembles that are probably neither beneficial nor harmful to the genes of most of their carriers. They can be seen as commensals that have colonized a special mental niche, namely the brain structures that more normally control curiosity and exploration, behaviours that contribute much to genome fitness.

Memes, due to the fact that they can manipulate the behaviour of their hosts, are predestined to increase their memetic fitness at a cost of some host biological fitness. They have it in their hands, as it were, to put their hosts' behaviour to work on their transmission rather than that of their hosts' genes, not unlike the rabies virus. Meme mutations analogous to parasites, that have high cultural fitness at the expense of host vigour, seem nearly inevitable. It is easy to see, for example, how memes inducing drug-taking get past gene-instructed censoring. Addictive drugs happen to activate the reward-signalling mechanism that is so important for learning, even though they are not in fact fitness-promoting, as are the stimuli that normally drive these reward circuits. In much the same way as saccharin fools the alimentary system, so does morphine fool the reinforcement system (Falk and Feingold, 1987).

Innate evaluation mechanisms in humans seem undiscerning about the optimal measure of resource seeking and holding efforts. Attempts to get the best out of the environment are certainly biologically advantageous to an individual, but at some point the returns cease to justify the investment. It is even suspected, for example, that birds occasionally miss mating opportunities on account of exaggerated territorial aggression. Among humans, greed often inhibits fertility. Many a successful parasitic meme profits from the quest for capital riches rather than for genetic fitness. Much of the commercial culture that pervades the civilized world is a certain consequence of this.

Celibacy is an obvious parasite meme that causes a reduction of host reproduction. It is part of a meme complex, a memome that manipulates the brains of hosts so that it reduced their sexual activity but increases instead their proselytizing behaviour, much as the rabies virus inhibits the reproductive behaviour of its host in favour of infective behaviour. Incidentally, certain organismic parasites go one step further and actually castrate their hosts as a means to increase their own fitness (Baudoin, 1975). Among Catholics the celibacy meme carried by one set of hosts is compensated by a linked meme expressing itself in the demand that the remainder of the memome carriers should commit themselves to relentless reproduction. On the other hand, competing memes that instruct the use of contraceptives have spread among Western cultures in recent years, to the extent that relevant populations are numerically decreasing. Contraceptives of course sever the innate link between reproduction and pleasure that normally ensures fertility.

The host's genome should in principle counteradapt against parasitic memes, but the faster evolutionary pace of memes versus genes makes that difficult. Moreover, selection for gene mutants against infection by specific memes can only become effective as these memes spread and are already part of the population's cultural heritage. As an analogy, a partial resistance against myxomatosis only began to emerge as a genetic trait among Australian rabbits after the disease had taken the character of a pandemic and the rabbits were close to extinction (Fenner and Ratcliffe, 1965). Still, gene mutations that somehow ensure brains immune to invasion by parasite memes are at an advantage against those that do not. However, meme mutations bypassing that resistance are culturally selected for, and so there is again a coevolutionary arms-race: hosts evolve improved censoring, memes evolve enhanced propagation.

This all suggests that in an advanced culture parasitic memes should be able to proliferate. Genes are unlikely to be able to devise innate defences against each and every one of the myriads of biologically harmful meme mutants that arise in as variegated a culture as the human one. One has even to consider the possibility that parasitic memes, such as those responsible for environmental pollution, could eradicate their human hosts, even before the genes of AIDS manage to do so.

CULTURAL EVOLUTION

Cultural evolution is the inevitable spin-off from the imperfect replicative properties of memes. Memes reproduce and mutate as they establish themselves in a given brain, and as they transmit themselves to other brains. The survival and the reproductive efficiency of different memes is not identical as they compete for and interact with their environmental niche. Some memes spread explosively, others are only mildly successful, while many become extinct. Different memes have differing cultural fitnesses, much as different genes have differing biological fitnesses. In short, the memetic information lodged in the collective memory of a given cultural ensemble is subject to variation and selection. Memes have to be viewed as independently evolving entities whose core habitat happens to be the brains of some higher animals and whose phenotypoic expression is the cultural behaviour of these same animals. In their essentials they are not too different from, for example, influenza viruses that inhabit the naso-oral cavities of vertebrates and express themselves in the sneezing and coughing behaviour of their hosts (Yamashita *et al.*, 1988).

The multitude of species and subspecies populating the earth is doubtlessly the most striking product of genetic evolution (Minkhoff, 1983; Futuyama, 1986). Speciation consists of the emergence of assemblies of mutually adjusted genes (genomes) adapted to survive and reproduce in different ecological niches. Subcultures and cultures can be similarly understood as distinct coadapted assemblies of memes, as populations of memomes, which thrive in different socioecological niches. An at least temporary isolation between pools of genes facilitates biological speciation. Restriction of meme flow for whatever reason, but often due to geographical separation between host populations, is an important factor in cultural speciation. Media and mobility are the antithesis of cultural speciation, as they facilitate the transport of memes between previously isolated cultures. The almost universal spreading of the Coca-Cola subculture in the late 1940s, and the MacDonald's subculture in the early 1980s, are witness to this. On the other hand, the tendency for like to mate with like, that is assortive mating according to characteristics such as height, eye colour, personality, etc., helps to maintain biological distinctiveness. The tendency for individuals of like culture to stick together, illustrated by both the isolation of immigrant communities and the insulation of social classes, in turn aids the preservation of cultural specificities.

When only a few individuals are the founders of a large population then the latter's genetic composition reflects its restricted ancestry. The analogue of this founder effect that favours the emergence of new species on islands also effects cultural evolution. It is known, for example, that only a few chaffinches colonized the Chatham Islands in the South Pacific in about 1900. The present population of this bird, some 35 generations hence, still has an aberrant and reduced song repertoire, a dialect that differs from that of the parent population living in New Zealand. No doubt this reflects the few and individaul song styles that the founders brought with them and passed on to their descendants (Baker and Jenkins, 1987).

Competition is a very salient characteristic of biological evolution. The replicative and instructive activities of genes are dependent on environmental resources. Finite resources limit reproduction and their partitioning leads to various forms of competition between genomes. In organisms capable of behaving, competitiveness frequently surfaces in the guise of agonistic behaviour. Aggression for food and space, strife about social rank and contests for sexual partners are examples. Memes also compete for limited resources, primarily for synaptic space in hosts, but also for the means that they need to reproduce themselves (principally a share of the hosts' behaviour). It is not surprising therefore that memes should also instruct their hosts to behave competitively, even agonistically on their behalf. Among humans at any rate, culturally driven aggressive behaviour is sadly often in evidence, even in its most extreme forms. Brawls among soccer fans, murder among political partisans, wars between religious sects, are events that challenge again and again our naive belief in human morality and rationality.

Biological evolution, however, also yields cooperative behaviour. Each member in a bird flock, for instance, benefits in fitness from the fact that the antipredator vigilance is enhanced by socializing: many eyes see more than two (Tinbergen, 1953). Culturally determined behaviour of this kind is extremely widespread, at least among humans. Religious sects, learned societies, political parties, etc., clearly arise because the memes concerned are more effective jointly than singly in spreading themselves.

The simple fact that blood relatives share varying proportions of genes generates selective pressure for the emergence of an accordingly graded altruistic disposition among them (Hamilton, 1964). Analogously, individuals can share many, few or no memes, yielding a gradation of memetic kinships. In advanced cultures with institutions such as schools and universities, large numbers of memes are transmitted among unrelated persons. Altruistic behaviour of a markedly parental quality by professors towards their 'best' pupils (i.e. those that have adopted many of their memes) is not uncommon. Culturally based helping behaviour among genetically unrelated people that have some beliefs or traditions in common is widespread: Muslims help Muslims, freemasons aid freemasons, fraternity members assist fraternity members, etc. In fact, cultural altruism may often simply reinforce or formalize the other biologically viable form of altruism, namely reciprocal altruism that, among animals at least relies on only a rather indefinite fellowship (Trivers, 1985).

Competition between biological kin groups can, however, also enhance strife. Capuchin monkey bands composed mainly of relations, for example, engage in quite warlike aggression against other bands about trees in fruit or stray females (Delius, unpublished observations). To an even greater extent, the same applies to cultural kin groups. Indeed, all too frequently Protestants and Catholics, Sikhs and Hindus and many other such groups choose to kill each other. Meme selfishness may on occasions even override gene altruism. Differing political allegiances for instance can make mortal foes of even close blood relatives, as documented by several tragic Spanish Civil War episodes.

Every culture seems to contain items that are in some way extravagant, involving effort, expense or inconvenience that appears disproportionate relative to the pay-off, return or advantage the items provide. Megalithic stone circles, pyramids, gothic cathedrals, tulipomania, operatic performances, fanciful fashions are some examples. Can these be compared to biological extravaganzas such as elk antlers, bird of paradise plumage, manakin dances or orchid flowers? In the biological context it is coevolution that most often brings about extraordinary traits. Whenever the evolution of two or more kinds of organisms is closely interdependent, in the sense that each kind is a selection agency affecting the evolution of the other, then there is scope for unpredictable, sometimes spectacular developments (Futayama and Slatkin, 1983).

Males and females of one species are often involved in such runaway games through sexual selection. Within each sex there is competition for the best sexual partners. Females, who invest heavily during reproduction, can gain much by choosing males with characteristics that promise offspring of quality. This generates selection for males that have such characteristics, but also leads to breeding females having ever stronger preferences. The ensuing feedback spiral can give rise to unusual features such as the peacock's tail or the bowerbird's bower (Borgia, 1986). Mutualistic memes that are like extensions of genes are bound to get caught up in this sort of game. The whole birdsong culture is strongly suspected to be a memetic offshoot of gene-based sexual selection (Catchpole, 1987). Among humans, sex-differentiated dress fashions have probably originated in the same way.

Meme reproduction itself at first sight seems to be of an asexual kind, much as the simple budding or cloning typical of such organisms as virus and bacteria (Jackson *et al.*, 1986). According to detailed questionnaire investigations by Cavalli-Sforza *et al.* (1982) on the cultural traits of American college students, the meme occasioning art museum visits for instance, appears to derive solely from the meme borne by just one other model person (father). However, the propagation of the meme motivating churchgoing among this population appears to require the fusion of appropriate memes carried by two persons (the parents). Other memes, for instance that eliciting jogging, usually descends from several memes born by friends, celebrities, etc., but not the parents. It was argued earlier that host genomes may in fact tend to bias memetic reproduction towards a multifusional mode involving several source memes ('do what everybody does'). It is uncertain, however, whether this primitive isogamic 'sexuality' of memes can support anything like cultural sexual selection.

Cultural trait luxuriation is more likely to be produced by the same kind of coevolutionary tangles effective in complex biological communities which occupy elaborate niches such as the humid tropics. There the survival and reproduction chances of any organism are mainly determined by the ecological context created by the other organisms, rather than by the physical conditions of the habitat. The intricate and dynamic web of organismal interactions characterizing such communities has led to the evolution of innumerable freaks, such as flowers that look like bees, caterpillars that look like snakes, moths that look like hummingbirds, butterflies that look like other butterflies, and so forth. Analogously, how well a meme succeeds depends largely on the cultural context in which it finds itself. For example, once a set of Muslim or Catholic memes has established itself in a brain, it generates a strong bias for the acceptance of further memes of Muslim or Catholic type, but also for the rejection of any Buddhist or Hindu memes. A runaway process leading to exaggeration and fanaticism becomes a strong possibility in such a context. From there to pyramid and cathedral building may be just a step further. More generally, the selection of memes by memes must doubtlessly be a major factor in cultural evolution. The arbitrary developments that such an inherently unstable arrangement can produce are intuitively boundless.

If memes select memes, can memes also select genes? In certain circumstances they undoubtedly do. Different socially learned songs, for example, definitely promote and sustain the genetical differences between several sympatric bird species (Thielcke, 1973). That is of course precisely what Tinbergen and Lorenz had assumed during that remarkable 1959 conversation mentioned at the beginning of this essay. Given that culturally transmitted song dialects also influence intraspecific mating preferences of songbirds, they are also bound to have subtle effects on their population genetics, even if the effects are difficult to pinpiont (Barker and Cunningham, 1985). During the Ravenglass discussions Niko Tinbergen often considered whether cultural processes such as language, birth control or modern medicine, were not importantly affecting the course of human biological evolution. A few years later he seemed to be certain about it. His worry was then that cultural evolution in several respects was outstripping biological evolution at a dangerously accelerating pace (Tinbergen, 1977). He wrote: 'It is an illusion to believe . . . that cultural evolution (entails) unmitigated progress'. The overall course of cultural evolution, he concluded, needed a rational and urgent correction. 'Time is running out', Niko warned us, and that is no less true today than it was then.

ACKNOWLEDGEMENTS

While preparing this chapter the author's research was supported by the Deutsche Forschungsgemeinschaft. J. Delius, J. Emmerton, M. Dawkins and R. Dawkins are thanked for critical comments and stylistic improvements. J. Delius also provided the Butler quote. M. Siemann and A. Niemuth assisted with patient and expert manuscript preparation. The chapter is an extensively reworked version of an earlier and longer text (Delius, 1989).

REFERENCES

Alderks, C.E. (1986) Observational learning in the pigeon: effects of model's rate of response and percentage of reinforcement. *Animal Learning Behaviour*, **14**, 331–35.

Anderson, R.M. and May, R.M. (eds) (1982) *Population biology of infectious diseases*, Springer, Berlin.

Baker, A.J. and Jenkins, P.F. (1987) Founder effect and cultural evolution of songs in an isolated population of chaffinches, *Fringilla coelebs*, in the Chatham Islands. *Animal Behaviour*, **35**, 1793–1803.

Bajema, C.J. (1972) Transmission of information about the environment in the human species: a cybernetic view of genetic and cultural evolution. *Social Biology*, **19**, 224–26.

Ball, I.A. (1984) Memes as replicators. *Ethology and Sociobiology*, **5**, 145–61.

Barker, M.C. and Cunningham, M.A. (1985) The biology of bird song dialects. *Behavioural and Brain Sciences*, **8**, 85–133.

Baudoin, M. (1975) Host castration as a parasitic strategy. *Evolution*, **29**, 335–52.

Biederman, G.B., Robertson, H.A. and Vaughan, M. (1986) Observational learning of two visual discriminations by pigeons: a within-subject design. *Journal for the Experimental Analysis of Behaviour*, **46**, 45–9.

Bonner, J.T. (1980) *The evolution of culture in animals*, Princeton University Press, Princeton.

Borgia, G. (1986) Sexual selection in bowerbirds. *Scientific American*, **256**, 70–79.

Boyd, R. and Richerson, P.J.V. (1985) *Culture and the evolutionary process*, Chicago University Press, Chicago.

Campbell, D.T. (1969) Variation and selective retention in sociocultural evolution. *General Systematics Yearbook*, **14**, 69–85.

Cann, R.L., Stoneking, M. and Wilson, A.C. (1987) Mitochondrial DNA and human evolution. *Nature*, **325**, 31–36.

Catania, C. (1985) Rule governed behaviour and the origins of language, in *Behaviour analysis and contemporary psychology*, (eds C.F. Lowe, M. Richelle, D.E. Blackman and C.M. Bradshaw), Erlbaum, London, pp. 135–156.

Catchpole, C.K. (1986). The biology and evolution of bird songs. *Perspectives in Biology and Medicine*, **30**, 47–62.

Catchpole, C.K. (1987) Bird song, sexual selection and female choice. *Trends in Ecology and Evolution*, **2**, 94–97.

Cavalli-Sforza, L.L. and Feldman, M.W. (1981) *Cultural transmission and evolution, a quantitative approach*, Princeton University Press, Princeton.

Cavalli-Sforza, L.L., Feldman, M.W., Chen, K.H. and Dornbusch, S.M. (1982) Theory and observation in cultural transmission. *Science* **218**, 19–27.

Chesler, P. (1969) Maternal influence in learning by observation in kittens. *Science* **166**, 901–903.

Changeaux, J.-P., Heideman, T. and Patte, P. 1984. Learning by selection, in *The biology of learning* (eds P. Marler and H.S. Terrace), Springer, Berlin, pp. 115–133.

Christie, A.B. (1981) Rabies. *Journal of Infection*, **3**, 201–218.

Curio, E., Ernst, E. and Vieth, W. (1978) Cultural transmission of enemy recognition: one function of mobbing. *Science*, **202**, 899–901.

Davies, B.D., Dulbecco, R., Eisen, H.N. and Ginsberg, H.S. (eds) (1980) *Microbiology*, Harper and Row, Philadelphia.

Dawkins, R. (1976) *The selfish gene*. Oxford University Press, Oxford

Delius, J.D. (1989) Of mind memes and brain bugs, a natural history of culture, in *The nature of culture*, (ed W.A. Koch), Brockmeyer, Bochum, pp. 26–79

Delius, J.D. (1990) Sapient sauropsids and hollering hominids, in *Geneses of language*, (ed W. Koch). Brockmeyer, Bochum, pp. 1–29.

Devoogd, T.J., Nixdorf,B. and Nottebohm, F. (1985) Synaptogenesis and changes in synaptic morphology related to acquisition of a new behaviour. *Brain Research*, **329**, 304–308.

Dudai, Y. (1989) *The neurobiology of memory: concepts, findings, trends*. Oxford University Press, Oxford.

Falk, J.L. and Feingold, D.A. (1987) Environmental and cultural factors in the behavioral action of drugs, in *Psychopharmacology: the third generation of progress*, (ed H.Y. Meltzer), Raven Press, New York, pp. 1503–1509.

Fenner, F. and Ratcliffe, F.N. (1965) *Myxomatosis*, Cambridge University Press, Cambridge.

Futuyma, D.J. (1986) *Evolutionary biology*, 2nd edn, Sinauer, Sunderland, Mass.

Futuyma, D.J. and Slatkin, M. (1983) *Coevolution*, Sinauer, Sunderland, Mass.

Ewer, R.F. (1969) The 'instinct to teach'. *Nature*, **222**, 698.

Gould, J.L. and Marler, P.(1987) Learning by instinct. *Scientific American*, **256**, 62–73.

Hamilton, W.D. (1964) The evolution of social behavior. *Journal of Theoretical Biology*, **7**, 1–52.

Harris, M. (1987) *Cultural anthropology*, Harper and Row, New York.

Hewlett, B.S. and Cavalli-Sforza, L.L. (1986) Cultural transmission among Aka pygmies. *American Anthropologist*, **88**, 922–34.

Hogan, D.E. (1986) Observational learning of a conditional hue discrimination in pigeons. *Learning and Motivation*, **17**, 40–58.

Hohorst, W. (1981) Parasitologie, in *Biologie*, (eds D. Starck, K. Fiedler, P. Harth and J. Richter), Verlag Chemie, Weinheim, S. 765–808.

Horn, G. (1986) Imprinting, learning, and memory. *Behavioral neuroscience*, **100**, 825–32.

Jackson, J.B.C., Buss, L.W. and Cook, R.E. (eds) (1986) *Population biology and evolution of clonal organisms*, Yale University Press, New Haven.

Klopfer, P.H. and Hailman, J.P. (1965) Habitat selection in birds. *Advances in the Study of Behavior*, **1**, 279–303.

Kohonen, T. (1984) *Self organization and associative memory*, Springer, Berlin.

Kroodsma, D.E. and Miller, E.H. (eds) (1982) *Acoustic communication in birds*, Academic Press, New York.

Lefebvre, L. (1986) Cultural diffusion of a novel food-finding behaviour in urban pigeons: an experimental field test. *Ethology*, **71**, 295–304.

Lumsden, C.J. and Wilson, E.O. (1981) *Genes, mind and culture, the coevolutionary process*, Havard University Press, Cambridge, Mass.

Margulis, L. (1981) *Symbiosis in cell evolution*, Freeman, San Fransisco.

Medawar, P.B. (1976) Does ethology throw any light on human behaviour? in *Growing points in ethology*, (eds P.P.G. Bateson and R.A. Hinde), Cambridge University Press, Cambridge, pp. 497–506.

Meire, P.M. and Ervynick, A. (1986) Are oystercatchers (*Haemotopus ostralegus*) selecting the most profitable mussels (*Mytilus edulis*)? *Animal Behavior*, **34**, 1427–35.

Minkhoff, E.C. (1983) *Evolutionary biology*, Addison Wesley, Reading, Mass.

Moore, J. (1984) Parasites that change the behaviour of their host. *Scientific American*, **250**, 82–89.

Morris, R.G.M., Kandel, E.R. and Squire, L.R. (1988) Learning and memory. *Trends in Neuroscience*, **11**, 125–79.

Murton, R.K., Coombs, C.F.B. and Thearle, R.J.P. (1972) Ecological studies of the feral pigeon. Flock behaviour and social organization. *Journal of Applied Ecology*, **9**, 875–89.

Nishida, T. (1987) Local traditions and cultural transmission, in *Primate societies* (eds B.B. Smuts, D.L. Cheney, R.M. Seyfarth, R.W. Wrangham, and T.T. Struhsaker), Chicago University Press, Chicago, pp. 462–474.

Norton-Griffiths, M. (1967) Some ecological aspects of feeding behaviour of the oystercatcher *Haematopus ostralegus* on the edible mussel *Mytilus edulis*. *Ibis*, **109**, 412–24.

Palm, G. (1982) *Associative memory*. Springer, Berlin.

Pringle, J.W.S. (1951) On the parallel between learning and evolution. *Behaviour*, **3**, 174–215.

Rummelhardt D.E. and McClelland, J.L. (1986) *Parallel distributed processing. Exploration in the microstructure of cognition*. MIT Press, Cambridge, Mass.

Schmidt-Koenig, K. (1965) Current problems in bird orientation. *Advances in the Study of Behavior*, **1**, 217–78.

Smith, D.C. and Douglas, A.E. (1987) *The biology of symbiosis*, Arnold, London.

Staddon, J.E.R. (1983) *Adaptive behaviour and learning*, Cambridge University Press, Cambridge.

Subowski, M.D. (1989) Recognition in ethology. *Perspectives in Ethology*, **8**, 137–71.

Thielke, G. (1973) *Die Wirkung erlernter Signale auf die Artbildung*, Universitätsverlag, Konstanz.

Tinbergen, N. (1953) *Social behaviour in animals*, Methuen, London.

Tinbergen, N. (1976) Ethology in a changing world, in *Growing Points in Ethology*, (eds P.P.G. Bateson and R.A. Hinde), Cambridge University Press, Cambridge, pp. 507–27.

Tinbergen, N. (1977) Time is running out, unpublished manuscript.

Trivers, R. (1985) *Social evolution*, Cummings, Menlo Park, CA.

Wallhäusser, E. and Scheich, H. (1987) Auditory imprinting leads to differential 2-deoxyglucose uptake and dendritic spine loss in the chick rostral forebrain. *Developmental Brain Research*, **31**, 29–44.

Yamashita, M., Krystal, M., Fitch, W.M. and Palese P. (1988) Influenza B virus evolution: co-circulating lineages and comparison of evolutionary patter with those of influenza A and C viruses. *Virology*, **163**, 112–22.

Zentall, T.R. and Galef, B.G. Jr (eds) (1988) *Social learning: psychological and biological perspectives*. Erlbaum, Hillsdale.

7

Niko Tinbergen, comparative studies and evolution

MICHAEL H. ROBINSON

INTRODUCTION

All sciences have an ontogeny, and ethology is no exception. When I joined the Animal Behaviour Group in 1963, Niko Tinbergen was talking about 'a science as young as ours' (Tinbergen, 1963). At that time I think it is fair to say that British ethology was in its adolescent phase. It had certainly passed youth, which, like senescence, is characterized by innocence. We were then full of adolescent passion and idealism, but perhaps lacking some degree of judgement and adult cynicism. We believed in an attainable state of revelation about the mysteries of behaviour, and that somehow we were standard-bearers. Much of the language of those days centred on the function and evolution of behaviour and in fact, Niko's 'Aims and Methods in Ethology' was a kind of state of the Union document that assessed progress, problems, and in some ways redefined the mission of Oxford ethology. It contained a reiteration of the definition of ethology as the biological study of behaviour and, importantly for my present purpose, put evolutionary studies as one of its four key components. If we were truly evolutionary ethologists at that time, then the comparative method was one of the main devices in our armoury. I shall herein attempt a somewhat idiosyncratic review to show why I think comparative studies are extraordinarily important.

COMPARATIVE STUDIES IN EVOLUTIONARY BIOLOGY

Before moving on to discuss the comparative method in ethology, it is worth stepping back to look at its role in evolutionary biology as a whole. (This is not the place to discuss the philosophy of science or such issues as whether biology has a special flavour in terms of methodology. Ernst Mayr (1982, 1988) is a good source of detailed and beautifully balanced treatments of the fundamental problems involved.) It is safe to say that there is absolutely no doubt that many evolutionary questions are not susceptible to experimental investigation. Darwin's entire approach to evolution was based on comparative studies. Even now the revisionists Eldredge and Gould have developed their punctuationist crusade entirely from comparing fossil assemblages, and thus from comparative studies (Gould, 1980; Gould and Eldredge, 1977; Eldredge, 1985a, b, 1989). Punctuationism could be derived only from the fossil record. This is obvious because fossil assemblages are our only basis for tracing patterns of change in the history of biological systems, but comparisons within living forms also yield important evolutionary insights. These are worth examining because behaviour, like soft parts, leaves no fossils.

Origins and functions

Comparisons between living forms - apart from behaviour - can involve morphology, anatomy, embryology, physiology, nutrition, biochemistry, genetics, immunology and so on. Comparisons can be used to elucidate a variety of biological problems. For me, some uses of the comparative method rank far ahead of others, in simple reflection of my preferences and addictions within the science of biology. First and foremost for me are the interlinked questions of *origins* and *functions*. These are the issues that Dawkins (1986) raised so elegantly and so eloquently in the 'The Blind Watchmaker'. Essentially, we naturalists burn to know how structures of marvellous perfection reach that perfect state. The nub of the problem is answering Paley's famous question, which *he* could only answer by postulating a grand designer of all life. To the evolutionary biologist the challenge is to discover, by comparison, what forms of intermediate structures could precede present specialized 'end' products. Important as is defining the predecessors, determining their presumptive adaptive function is equally important. Niko (1959a), with characteristic clarity, called the entire two-part process 'the descriptive reconstruction of the course of evolution that has led to the present situation'.

Since becoming a zoo director, I have learned of a gloriously Kiplingesque example of evolutionary fine-tuning. It concerns the hairs on the tail of the pygmy hippopotamus. These are aborescent, the only mammalian hairs to branch, and their function involves territorial marking (Kranz, 1982). Hippos mark their territories with faeces which they scatter by whizzing their tails around while defaecating. Arborescent hairs fling faeces more efficiently than straight ones. How could such hairs evolve? What a confounding question. If we are to dismiss grand design, we must recognize that there are two questions involved in the answer to Paley: what are the steps in the process, and what was the function of the structures as they evolved? Comparative studies are the only source of answers.

Relationships

Of course, the most obvious use of comparative studies for the great majority of biologists, i.e. those not obsessed by the origin of intricate mechanisms/devices/structures/characters, is in elucidating relationships. Building grand phylogenies admittedly is fun; building accounts of relationships at the smaller scale of genera, families, orders and so on, can be immensely satisfying. Whether we do it by external characters - the preponderant method, because eyes and microscope-aided eyes were the only technologies available to Linnaeus and most of his descendants - or whether we now plug in chemical technologies, comparisons are paramount. They can be statistically based, intuitive, or both, but most frequently they are built on evolutionary interpretations. This is not always true. There are taxonomists who study patterns of resemblance in their own right, without assuming any evolutionary implications. The term that Dawkins (Dawkins, 1986) uses for this trend, 'pure resemblance measurers' should, I hope, stick.

Convergences, parallelisms, homologies and analogies

Comparative studies are also the basis of the study of the convergences and parallelisms which enrich evolutionary theory. Homologies and analogies also emerge from this field and can provide a high degree of intellectual excitement. A simple-to-describe example concerns the evolution of the camera-like eye convergently in the cephalopods and the vertebrates. As I watch our nautiloids swim clumsily around the National Zoo's Invertebrate Exhibit I realize that their very conservative design, dating back to the Mesozoic, still works. They also show us how an ectodermal pit could become an eye. One could apply the same process to convergent eyes in polyphyletic arthropods. For me as an airplane enthusiast, the 'discovery' of lift in the 'wings' of rays and skates, and its utilization for flight in insects, reptiles, birds and mammals raises the question of whether the Wright brothers (and other claimants) would have thought flight possible without the example of birds. Would they have developed the physics quicker if they'd studied the flight of non-buoyant fishes, or would they have arrived at the helicopter before the airplane if they had studied the flight of the dipterocarp fruits?

Evolutionary insights

Whether one believes that biologists proceed by induction or by the so-called hypothetico-deductive method (or both) there is little doubt that, as Niko so often insisted, asking the right questions needs a background of observations. The observations do not need to be one's own, but a broad knowledge of animal biology, derived from comparative studies, can certainly provide the insights necessary for hypothesis formation. It can provide insights into the existence of selection pressures both broad and narrow. In a sense, as Roberta Rubinoff pointed out to me, some kinds of comparison enable one to utilize data from the experiments carried out by the effects of stochastic natural processes, geological history, and evolution itself. Darwin's insights, derived from studying the Galapogos flora and fauna, were in fact derived from the results of a *long-term exclusion experiment*. The 'chance' geographical separation of the islands from the mainland produced the kind of situation that a present-day ornithologist might wish to contrive, one where he would be able to seed uninhabited areas with one family of birds, and exclude their competitors. Of course he would want to live long enough to observe the results. Darwin had the insight to interpret the results of this experiment partly because he had the comparative base acquired by a good and well-travelled naturalist. In his narrative of the Beagle voyage he remarked: 'Reviewing the facts here given, one is astonished at the amount of creative force, if such an expression may be used, displayed here on these small barren islands'. Australia is a great experiment for the marsupials, South America for the edentates, and so on. There are also great 'experiments' that result from closely related species occupying very different habitats.

COMPARATIVE STUDIES IN ETHOLOGY

Niko's attitude to comparative studies was itself evolutionary. It is interesting that in the 'Authors Notes' to his first volume of 'The animal in its world' (1972) he says 'It was Konrad Lorenz's relentless prodding which made me turn to comparative studies. Having studied the herring gull in some depth, I naturally turned (finally convinced by Lorenz's paper on dabbling ducks) to other gulls'. Of his (1959b) paper on comparative studies in gulls he says, 'it has a dimension hardly found in my earlier papers: inter-species comparison. More precisely, what was lacking in all those *one-species papers* was any serious attempt to apply the comparative method for the purpose of interpreting diversity and similarity as the outcome of adaptive divergence and convergence; as such they show a failure to take in the lessons I could have learned from Whitman, Heinroth, Huxley, Verwey and Lorenz' (my italics). This is a very frank evaluation. The fact is that Niko himself never carried out a broad comparative study, but his students certainly carried that banner for him. From cichlid fishes, Baerends and Baerends (1950), to cephalopods (Moynihan, 1985) and beyond, the list is long and honourable. My close association with Niko's pupil Martin Moynihan, during 20 years based in the tropics, has convinced me that he is *primus inter pares* in this field. Most of the later examples that I use are from his work.

None of this means that Niko did not realize the importance of comparative studies in ethology. 'The study of instinct' (1951) contains a major section on evolution and some provocative thinking on homologization and comparison. From that point on, regularly in the 1950s and 1960s Niko produced a succession of papers on behaviour evolution (1954, 1959a, 1959b, 1960, 1962, 1963, 1964, 1965a, 1965b, Hinde and Tinbergen, 1958). Many of these were concerned with experimental investigations of the survival of behaviour patterns, but some were treatments of the entire field and the role of comparative studies in the reconstruction of evolutionary pathways, in systematics, and in the elucidation of function and causation. Noteworthy are Hinde and Tinbergen (1958) and Tinbergen (1959a). In Hinde and Tinbergen (ibid) the authors define the limits and problems of the comparative method. They mention 39 comparative studies, then either completed or in progress, spanning spiders, grasshoppers, mantids, flies, wasps, fiddler crabs, cichlid fishes, sticklebacks, three families of finches, buntings and gulls. This may have been a fairly comprehensive list for the time. Anticipating the future emphasis of much of the Oxford group's research is an important paragraph: 'A knowledge of the probable course of evolution prompts further enquiry as to why evolution has taken that course and not some other. It is thus desirable to know whether the changes are adaptive and can have been brought about by selection. This involves a study of the survival value of the behaviour elements . . .'. In his 1959 paper 'Behaviour, systematics and natural selection' Niko examined the subject critically, giving some important recipes for using the comparative method. These are well worth rereading, although they may now be largely subsumed in more recent treatments of comparative methodology (Ridley, 1983). Distilling this original wisdom, and other protocols from Hinde and Tinbergen (ibid) we get:

1. 'As in morphology, successful use of the comparative method depends on the selection of characters to be compared'. (H and T)
2. 'It is not desirable to use characters which change rapidly and could have been acquired independently in different groups'. (H and T)
3. 'Many of the difficulties involved in the use of behavioural characters can be avoided by a broad approach; the importance of a knowledge of the natural history of the animal and the causation and function of the behaviour cannot be over-emphasized'. (H and T)
4. 'In order to translate differences between contemporary forms into changes of time, one step further is required; an interpretation of the direction of change. To do this it is essential as in comparative anatomy to combine data derived from the comparison of species with data about the function of both the original and the derived character'. (T)
5. 'The most convincing examples of evolutionary changes of function have been found in signal function'. (T)
6. The two most fruitful types of approach are (a) the study of divergent forms derived from common stock, and (b) that of distantly related convergent forms.' (T)

This is a selection that one can evaluate 30 years later. Tinbergen's treatment (1959a) also contains a crucial section on 'Interaction of selection pressures'. This is an idea he was to return to again and again. In essence he argued that behaviours may be evolutionary compromises. Thus when one is studying selection pressures, 'one cannot help seeing that selection pressures must often be in conflict with each other. One ends up by discovering that each character has not only been improved with regard to its own particular function but has undergone indirect effects which . . . are recognized as effects of selection when seen as parts of a system.'

At this stage it is worth giving some examples that I have selected principally from my own work and studies by my colleagues at the Smithsonian Tropical Research Institute. I made this choice mainly for two reasons, quite apart from the intrinsic content of the studies. First, I am more familiar with this research than with work elsewhere (my ten-year stint as an administrator 1980–1990 having isolated me from comparative research). Second, I think that they make points about tropical research compared with the state of university-based science. Tropical biology is uniquely rich in opportunities for comparative studies because there almost certainly are, for many major groups of animals, more related species present in any tropical area than anywhere else on earth. This is true of both rainforest faunas and those of coral reefs. This makes it easier to study very large species-samples in relatively small areas. In this respect, even zoos do not come near to the species richness of, say, Barro Colorado Island. The other factor relates to ethos and style. Research institutes can, if properly led, avoid the almost inevitable 'Rat-race effect' found at universities. In modern academia the probability of commitment to broad lateral studies, undertaken over long periods, is frequently low. This kind of commitment is essential for most comparative studies. (The universities most particularly affected by rat-racery are those where conditions of tenure involve assessements of paper productivity.) There is yet another reason for choosing some of these studies to exemplify the comparative approach: many were published in journals not necessarily read assiduously by ethologists. Criticisms of the conduct and results of comparative studies will be raised later.

DESCRIPTIVE RECONSTRUCTIONS OF THE COURSE OF BEHAVIOUR EVOLUTION

Camouflage, mimicry and defensive adaptations

When I joined the Behaviour Group, David Blest had just completed his study of the function of eye-spot patterns in Lepidoptera (1957a). This led him to some broad scale comparative studies, and the consequent evolutionary insights (1957b; 1963a; 1963b). Investigations of defensive systems in insects were not entirely a new venture for Niko's students. De Ruiter (1952; 1955) had previously studied counter-shading and stick-mimicry in caterpillars. Oxford in the days of Hope Professor E.B. Poulton had a long tradition (papers from 1884–1934!) of camouflage studies and Niko had worked with Bernard Kettlewell in experiments on industrial melanism. Later there was to be an extensive investigation of camouflage in the eggs of black-headed gulls (Tinbergen *et al.*, 1962). I started a study of the adaptive function of rocking movements in stick- and leaf-mimicking insects which was to lead me into a lifelong interest in predator–prey interactions. This thesis topic, with great serendipity, led me to Panama, where the diversity of insects is overwhelming. This was long before the studies of Erwin (1982) suggested that there were perhaps 30 million species of insects in the world, most of them in the rainforests, rather than the 1.5 million previously estimated. In Panama, and later in Colombia, Venezuela, Sri Lanka and Papua New Guinea I studied the defensive adaptations of phasmids, mantids, tettigoniids, acridiids, and, eventually, spiders. In these groups was a whole range of visual defences, ranging from relatively simple camouflage to the most complex and detailed forms of stick, leaf and faeces mimicry. As I accumulated examples of the complex postures assumed by both eucryptic and mimetic insects (Robinson, 1964; 1968a, b, c, d; 1969a, b,; 1970a, b; 1973; 1977; 1978; 1979; 1981; 1985; Robinson and Pratt, 1976), the comparisons across groups and between groups, together with the mine of information contained in the comparative studies of mantids by Crane (1952) for Venezuela, and Edmunds (1972, 1976) for West Africa, eventually clicked together. I was led to visualize a scheme that accounted for the evolution of stick mimicry and, coincidentally, raised hypotheses about the prey-recognition processes of sophisticated visually hunting predators (1969a; 1970b; 1985). If one wishes to reconstruct possible steps in the evolution of specializations, one needs to discover a good range of probable intermediates and also to find adequate inferences or evidence about the selection pressures involved in the changes. From my own work and a wide variety of other studies I eventually had both of these essential elements. As my experience of diverse tropical faunas has expanded, convergences between unrelated groups, and between the same groups in widely separated geographic regions seem now to fit within the original scheme and pile more and more examples into it (Figures 7.1 and 7.2).

The insight that proved to be a turning point came from the discovery of a remarkable stick insect in Panama. This was *Pterinoxylus spinulosus*, which has an unusually specialized resting posture (Robinson, 1968c), shown in detail in Figure 7.3. Not only are the first pair of legs held in line with the body axis and apposed, as in all stick-like phasmids, but the second and third pairs of legs are held in complex postures, folded on themselves in highly specialized ways. The immediate inference is that these are *leg-concealing postures*. The reciprocal conclusion is that such devices imply that predators may attend to such giveaway features of insect prey as the possession of legs.

Once one assumes that legs may be a disadvantage to concealment, disguise and deception, then each new species of camouflaged or mimetic animal one finds becomes a source of anticipation. Added to the excitement of collection and discovery (perhaps an atavism from hunting–gathering) is a new curiosity based on the question 'what will it do with its legs?' Despite the fact that fewer than 5% of the

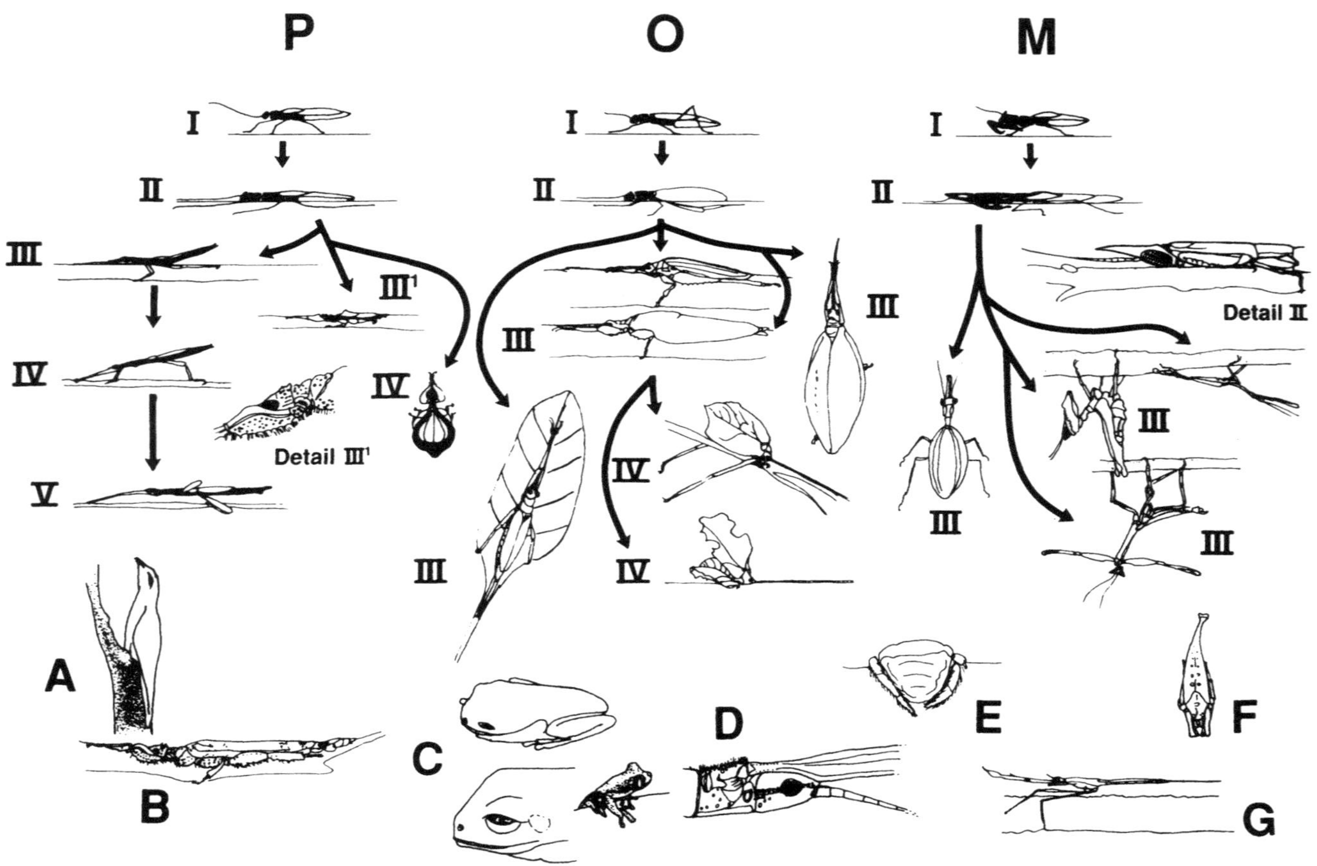

P
O
M
I
II
III
III¹
IV
V
Detail III¹
Detail II
A
B
C
D
E
F
G

Figure 7.1 The evolution of plant-mimicry in insects with some examples from other groups. This diagram shows hypothetical steps in the evolution of stick and leaf mimicry, principally in three orders of insects. It is derived from studies cited in the text. The three lines shown are grouped under P for Phasmida, O for Orthoptera, M for Mantids, sometimes regarded as a subgroup of the Dictyoptera. In each line, stage I is the generalized ancestral form, free-standing and possessing generalized, colour-matching camouflage. Stage II is still camouflaged but rests prone against an appropriate substrate and conceals typical insect structures by behavioural processes that produce either elongation, leading to stick mimicry, or flattening, leading to leaf mimicry. In the phasmid line morphological elongation in specialized forms results from hypertropy of the mesothorax, and behavioural postures (stages III, IV, and V). In the orthopteran line leaf mimicry is achieved by various modification of the forewings, and enhanced by postures (stages III, and IV). In the mantid line complex leg postures enhance apparent length and supplement pro- and mesothorax elongation. Details of the genera and species diagrammed here are to be found in the references cited in the text. The supplementary illustrations A to G illustrate some other mimetic posture. A is the Caprimulgiform bird, the Potoo, that rests in a branch-like posture on trees. Note the closed eyes. B is the phasmid *Prisopus berosus*, stage III, which is flattened and concave ventrally. It rests with its ventral surface tightly appressed to twigs. This flattening could be the precursor of the marvellous leaf mimicry seen in *Phyllium* species. C is the red-eyed tree frog *Agelychnis callidryas*, which assumes a cryptic posture in which the legs are concealed, as are the bright red eyes. In the non-mimetic posture the frog reveals conspicuous red–orange markings on the inside faces of all four legs. D shows the notched leg-bases of all stick-mimicking phasmids. These facilitate protraction of the legs in line with the body axis and enhance the stick mimicry. E is the araneid spider *Caerostris tuberculata* that assumes a diurnal cryptic posture on twigs, concealing all its legs by apposition to the body, as in the insect lines P,O,M, II. F is the flower-mimicking spider *Arachnura melanura* which also apposes the legs. G is the water stick-insect *Ranatra*, mentioned in the text (after Cloarc, 1988). This stick posture is exactly parallel with that of the totally unrelated mantid shown at the extreme left of the MIII stage above.

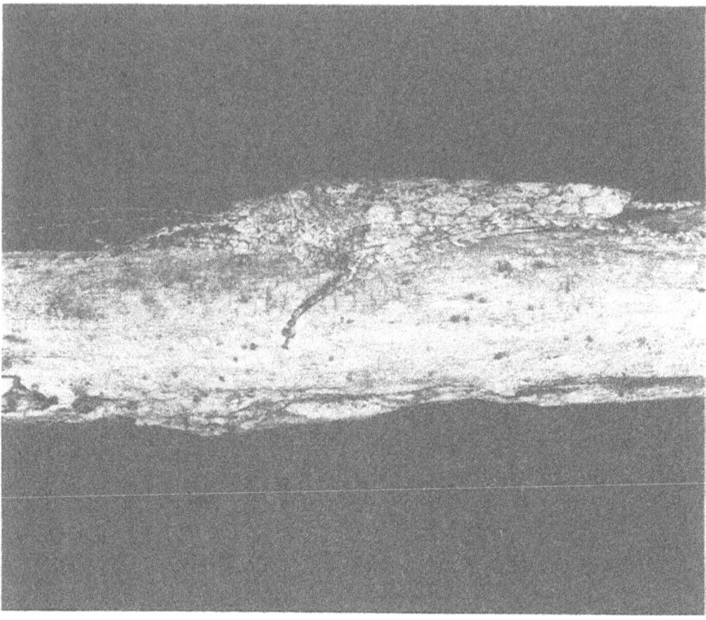

Figure 7.2 Photographs of the dead-leaf mantid *Acanthops falcata* and the prostrate-resting cryptic tettigoniid *Acanthodis curvidens*, shown in Figure 7.1 stages MIII and OII respectively. Note the postures of legs I.

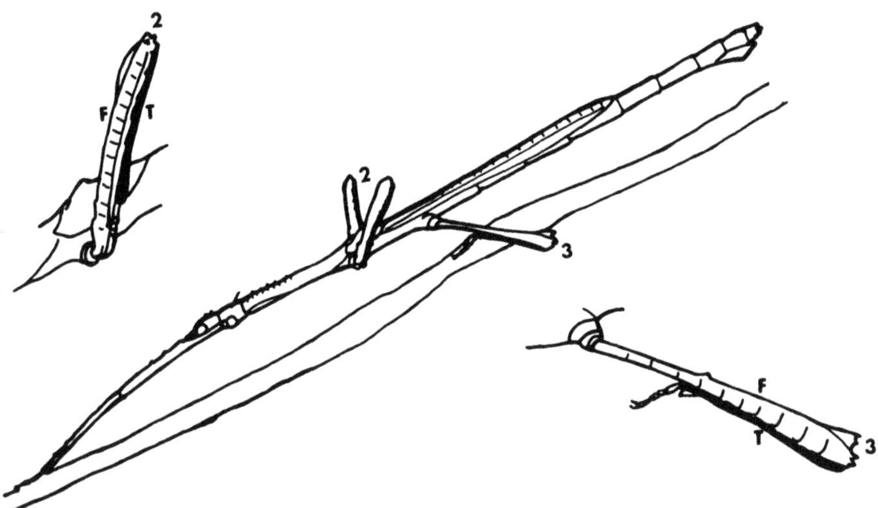

Figure 7.3 The phasmid *Pterinoxylus spinulosus* referred to in the text (page 107). This insect assumes a posture that effectively conceals all six legs in positions consistent with the overall stick mimicry.

living insects I collected could be named, I was never disappointed. Their behaviours were a revelation: one was constantly amazed by the variety of concealing 'inventions' involved. Instances of these have accumulated with time (Robinson, 1970b, 1985). Clues come from other people's photographs, where the significance of the postures they have recorded had not been recognized, at least overtly. For instance, quite by chance while surveying the recent behaviour literature for this paper, I came upon a drawing (Cloarec, 1988) of the mimetic posture of the water stick-insect *Ranatra* sp., a hemipteran, far indeed from the phasmids. This posture is strikingly similar to that of a praying mantis from Panama that I described over 20 years ago (Robinson, 1969b). One of the most satisfying things for me as a naturalist, despite the absence of taxonomists capable of identifying the beasts, was to find three species of elusive leaf insect (*Phyllidae*) in Papua New Guinea, each of which folded its legs to merge totally with its body. Things illustrated in classic books are a glory to see in the wild.

The use of the comparative method to identify widespread selection pressures for concealing characteristic insect parts eventually led to the final satisfaction of being able to propose a reconstruction of how stick mimicry could have been evolved. This was made easier by the fact that both the mantids and the phasmids (two widely mimetic groups of insects) had present-day forms that seemed likely to be similar to probable stages in the evolutionary sequence leading from camouflage to stick or leaf mimicry. My reconstruction started from insights gained from studying camouflaged insects resting in twigs and leaves. These align themselves along the twig or leaf vein (in Papua New Guinea, along the concave upper surfaces of *Pandannis sp*. leaves). In their cryptic postures they protract their anterior legs in line with the long axis, fold their intermediate legs against the body or substrate, and position the posterior legs behind the body in line with its axis (Robinson, 1969b). This posture effectively conceals the legs, antennae and head, and reduces the profile of the insect. It also, in effect, increases the insect's length by adding the length of legs I and III to the body length. Of course, if the camouflage works, none of this extension is apparent. However, the postures can be regarded as a step towards potential mimicry. There are several phasmids that rest in this position. They are simply cryptic, background-matching species (Robinson, 1969b). If the insect does not choose to rest on a background appropriate to its body colour it could be at risk, or at least provoke further scrutiny. Any cryptic insect resting on an elongate structure could, from this posture, evolve toward further profile reduction (concealment) as a result of further elongation (Figures 7.1, 7.2). Interestingly enough there are two other solutions to the leg concealment problem. Both are 'dead ends' that cannot lead directly to stick mimicry. One leg-concealing variant is to tuck the hind legs under wing covers. This solution is accomplished in two different ways, depending on wing posture, seen in two genera of New Guinea katydids (Robinson, 1973). Yet another variant is to *fold* legs II and III against the body, a device found in New World tettigoniids of the family *Pseudophyllidae* (Robinson, 1969a).

The possible steps from an elongate, prone resting posture are shown in Figures 7.1 and 7.2. They are represented in the mimetic postures of existing species. To reiterate, this hypothetical pathway involves the assumption that legs and other characteristic animal structures are concealed because they are giveaway clues to the detection of otherwise concealed or disguised animals. The primary reason for assuming this is that there is an amazing range of behaviours and structures that seem to conceal legs and other structures. These adaptations are found in spiders of at least two families (Robinson, 1985, and unpublished data), lizards and even birds. Some of these are shown as adjuncts to Figure 7.1 in Figure 7.2. Despite the range and complexity of such presumptive leg concealment, the argument may seem to be circular: legs are concealed therefore legs are prey-detection clues. I believe that what this really constitutes is not circularity but the inductive base for a testable hypothesis. We can simply ask: do visually hunting predators actually use legs to detect prey? I have some very suggestive results from experiments on this problem (Robinson, 1970b). Among these results is the fact that sticks with cricket legs glued onto them were attacked and bitten by tamarins; similar sticks without legs were ignored.

A detailed study of arthropod antipredator adaptations leaves one with an indelible picture of a complex arms race between predators and prey. The detailed perfection of defensive devices delighted Niko over and over again. He argued strongly against the prevalent opinions that many defensive devices were too good to be true, or that in some sense they were better than necessary to fool the perception of predators (Tinbergen, 1963). I think he was right to do this. For me the complex perfection of defences seems to argue against the kinds of rapid macro-evolutionary change implied in some treatments of punctuationism.

Behavioural evolution in orb-weaving spiders

Araneid spiders are a very numerous part of the arthropod faunas of tropical rainforests (Elton, 1973). A considerable body of research has accumulated on their behaviour in the last 25 years (see for instance, Shear, 1986). Students have concentrated on predatory behaviour, courtship and mating behaviour, web structure and web building. These studies have led to reconstructions of evolutionary pathways and to the construction of phylogenies. My own studies of predatory behaviour, and those of Eberhard, and our numerous colleagues, co-workers and students have allowed us to reconstruct possible stages in the evolution of the complex attack and immobilization elements of araneid predatory behaviour. These studies have also provided us with functional explanations of the reconstructions. Robinson (1975) gives an overview of these reconstructions and the evidence for them. Later enhancements of that review can be found in Coddington (1986), Eberhard (1986) and Stowe (1986).

The focus of this story is on prey-wrapping behaviour by which orb-weavers enswathe their prey in silk. Eberhard (1967) drew attention to the fact that not all web-building spiders (a term that includes several families, not only the dominant orb-weavers) were capable of using silk at the first stage of their attack. He proposed a scheme based on comparisons between families to account for the origins of attack-wrapping. Later, I showed that all the stages for a hypothetical sequence of the evolution of attack wrapping exist within the one most diverse family of web-builders: the Araneidae (Robinson, 1969c). This scheme, shown in Figure 7.4, is supplemented by details of the

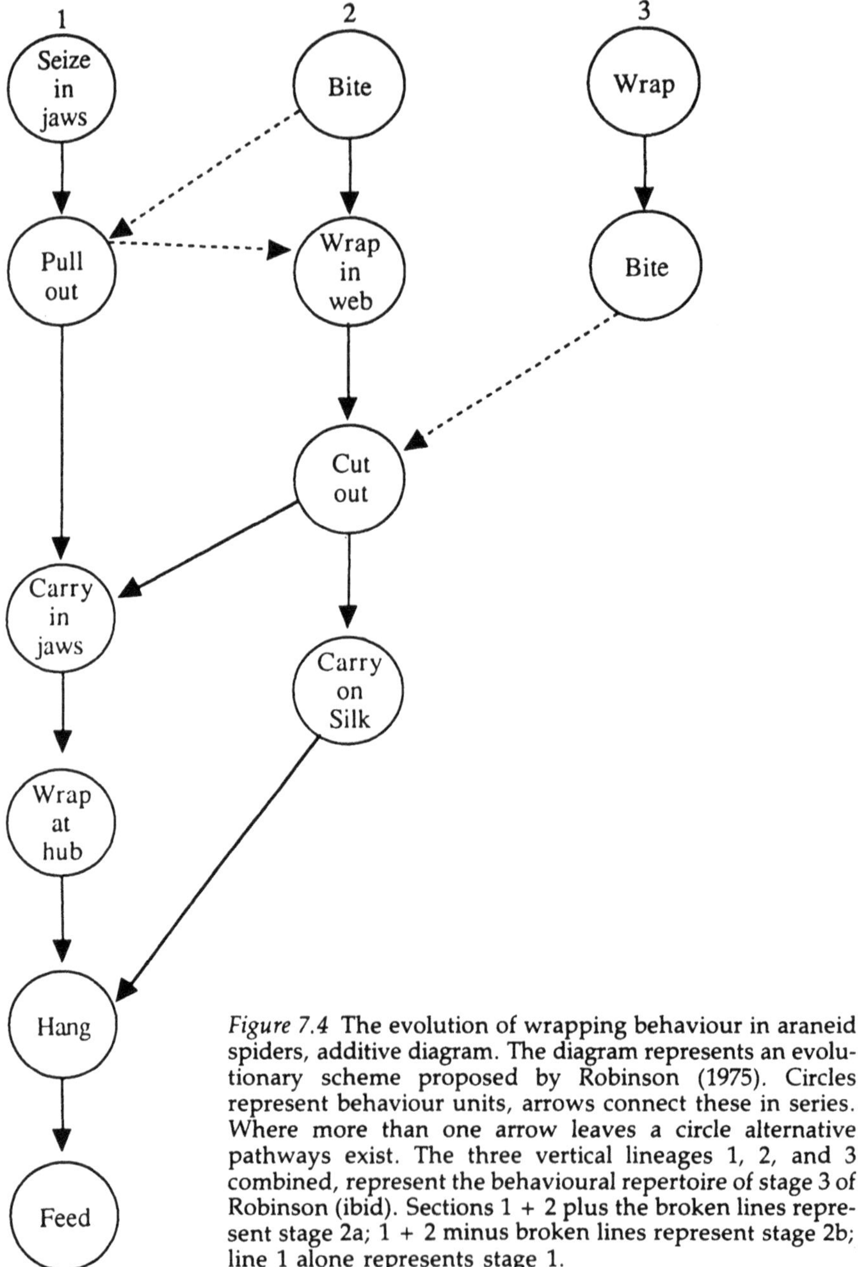

Figure 7.4 The evolution of wrapping behaviour in araneid spiders, additive diagram. The diagram represents an evolutionary scheme proposed by Robinson (1975). Circles represent behaviour units, arrows connect these in series. Where more than one arrow leaves a circle alternative pathways exist. The three vertical lineages 1, 2, and 3 combined, represent the behavioural repertoire of stage 3 of Robinson (ibid). Sections 1 + 2 plus the broken lines represent stage 2a; 1 + 2 minus broken lines represent stage 2b; line 1 alone represents stage 1.

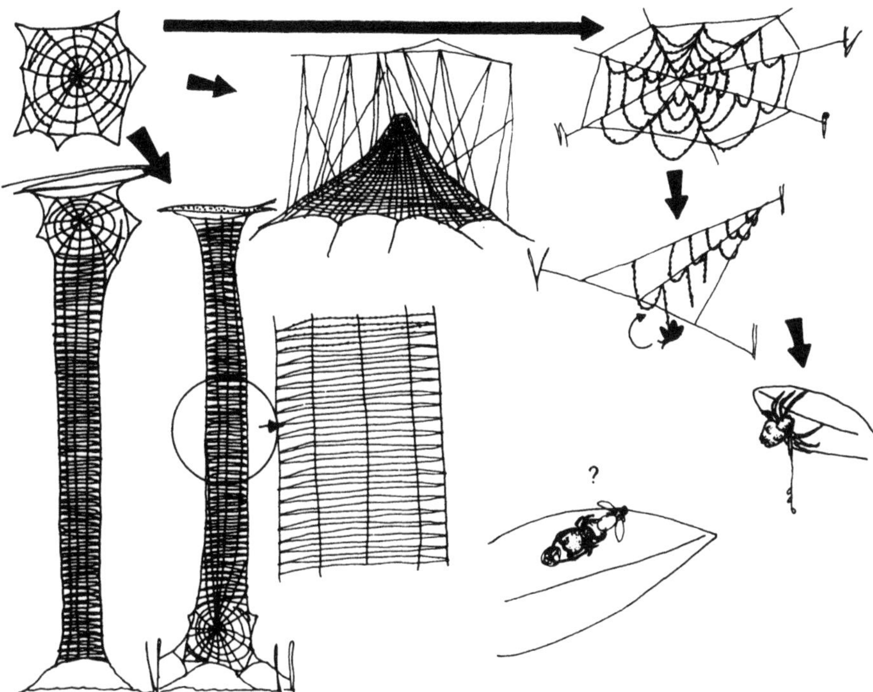

Figure 7.5 Evolution of orb-web structures. Arrows show the direction of specialization. *Upper left*: regular, more or less symmetrical orb. *Lower left*: ladder webs from New Guinea and Colombia. The detail shows the ladderlike viscid element. *Upper right*: horizontal orb of *Poecilopachys sp.* from Australia and New Guinea. The viscid elements are slack and hang down in loops; when an insect strikes one of them it shears at one end and absorbs the impact. Below this is shown the *Pasilobus* web, an orb reduced to one sector that functions in the same way. Bottom right: the bolas spiders *Mastophora* and *Ordgarius* make a single thread with one or two drops of viscid material (see text, pages 113–16).

functional advantage of each stage in such a reconstructed evolutionary pathway. The studies even included experimental evidence of the advantage of each addition to the 'primitive' state. Eberhard's recent (1982) phylogenies for the family araneidae emphasize the appearance of attack wrapping within the family.

Perhaps even more insight-provoking for the process of reconstructing pathways of web-evolution are discoveries about specializations of the originally wheel-like orb web. Web structure is an expression of behaviour and can be regarded as an expression of the evolution of species-specific behaviour. In 1970, with B. Robinson I discovered a remarkable specialization of the conventional orb in Papua

New Guinea (Robinson and Robinson, 1972). We called this a ladder web because it consisted of a parallel-sided extension of a lower segment of an orb in which the viscid element, spiral in an orb, was side-to-side like the rungs of a ladder. Eberhard then discovered an equally specialized 'inverted' ladder web in Colombia (Eberhard, 1975). This remarkable piece of convergence (Figure 7.5) between two unrelated spiders, separated by more than 15 000 miles, is attributable to the operation of a common selective factor. It is part of the continuing arms race between spiders and moths. Moths (and other lepidopterans) can escape readily from spiders webs (Eisner *et al.*, 1964, Robinson, 1969d). The scales on their wings come off onto the glue on the web, inactivate it, and they flutter free. The ladder web gives an exaggerated vertical extent of sticky surface to flutter against.

Eberhard (1977, 1980a) worked out the predatory mechanisms involved in the moth-catching behaviour of the bolas spider *Mastophora dizzydeani*, an araneid that uses as a capture device a single short thread with a terminal glue droplet. The moth secretes a chemical lure to attract male moths, and has almost certainly evolved a superglue. This device begs the question, what are the stages between the ancestral orb and this highly reduced silk structure? The discovery in Papua New Guinea of the *Pasilobus* web (Robinson and Robinson, 1975) provided both a link in the evolutionary sequence and the basis for reinterpreting the web of *Poecilopachys* described by Clyne (1973). The webs of both of these spiders are more or less horizontal and the viscid elements hang below the structural frame. Each sticky element has a low shear joint at one end, and this breaks on impact to absorb the shearing force exerted by the flying moth. The tethered insect is then killed by the spider. The *Poecilopachys* web is a complete orb; that of *Pasilobus* consists of only two sectors of an orb. The step from one to the other is one of reduction and concentration. The *Pasilobus* web can give rise to the bolas in much simpler steps than can the conventional orb. Figure 7.5 shows these structures. Further intermediates could await discovery.

Robinson and Robinson (1978, 1980) carried out a study of the courtship and mating behaviour of more than 50 species of araneid spiders. These studies revealed that courtship fell into three major categories according to both the position of the courting male in relation to the female's web, and to the type of primary sensory modalities involved. They regarded these categories as representing an evolutionary progression from primitive to advanced, but Eberhard (1982) disagreed. There *is* a correspondence between classification based on predatory behaviour and that based on courtship.

Behavioural comparisons and taxonomic relationships

Moynihan has contributed consistently to the elucidation of relationships through behaviour comparisons. In 1959 he produced a revision of the family Laridae; then in 1962 a major review of South American and Pacific larids. In that first major treatment Moynihan discussed the kinds of behaviour that provide 'reliable evidence of relationships', arguing that displays serve that function best. He defined display as 'any and all behaviour patterns that have become standardized or stereotyped in any way, in order to subserve a social signal function.' The usefulness of displays in taxonomy is, Moynihan argued, 'due to the fact that they are less often or less thoroughly, affected by convergent evolution than many morphological features' and 'displays need not be so closely adapted to as many aspects of the external environment'. Later he was to add (Moynihan, 1975) 'the physical features which are characteristic of most displays, that is stereotypy and exaggeration of form, are equally characteristic of some behaviour patterns which are adapted to promote crypsis'.

It seems to me that these judgements still stand: displays are good characters. However, good taxonomic judgements can be derived from other types of behaviour, as Eberhard has shown. After a really extensive series of investigations into web-building behaviour, Eberhard (1982) combined data from these studies with extensive data from his own and other studies of predatory behaviour and courtship behaviour to use 'behavioural characters for the higher classification of orb-weaving spiders'. He used data from 148 species, representing 'at least' 55 genera, to develop a distribution of conservative characters that agrees 'in general' with classical taxonomic schemes based on adult morphology. Eberhard concentrated his web-building studies on four phases of constructional behaviour. These were:

1. The construction of the radii (four distinct methods).
2. The construction of the temporary spiral of the web (three different styles).
3. The construction of the sticky spiral of the web (five distinct elements together totalling 13 different behaviours).
4. Construction of the hub of the web (five distinct behaviours).

This kind of behavioural analysis gives a lot of information to correlate. To illustrate the complexity of the behaviours, I can simplify just one of the sections of analysis, that on sticky spiral construction. To add the spiral, all orb-weavers work from the outside of the web inwards, after building the non-sticky structural scaffolding of the web. Most spiders construct a temporary spiral of dry silk from the inside outwards before laying down the sticky spiral. They use this as a foothold during their construction walk. As the spider moves from radius (spoke) to radius it attaches the spiral. How the spider gauges the attachment point is subject to very marked behavioural differences between genera. There are four different and distinct behaviours at this stage which are complex and difficult to describe in words alone. However, to describe just part of the movement illustrates the difference. In version (a) the spider reaches *forwards* with its *left first* leg; in version (b) it reaches *forwards* with its *right first* leg; in version (c) it reaches *back* with its *left fourth* leg, and in version (d) there is no reaching behaviour at all. There are also behaviour differences in which legs hold the radius just before the moment of attachment, and in contacts with the temporary spiral. This is probably as complex a comparative analysis of behaviour as has ever been documented. It revealed a number of convergences as well as illuminating relationships.

Comparative studies and the production of evolutionary insights

Moynihan's work on communication in squids, cuttlefish, nautiloids and octopus (1975; 1983a, b; 1985; Moynihan and Rodaniche, 1977; 1982) follows his long series of other comparative studies that has greatly illuminated several dark corners. Moynihan's work has ranged (in at least 25 substantial papers from 1955 to the present) from gulls, through birds of several other families, to New World primates, cephalopods and back to birds again. To me the most exciting aspect of the cephalopod work, out of many exciting features, is Moynihan's identification of the conservatism of certain displays. The cephalopods have a very ancient history and they are quite well represented by fossils, so we know when certain major groups separated: plus or minus a few million years, the three dominant orders of modern cephalopods Teuthida (squids), Sepiida (cuttlefish), Octopida (octopus and argonauts) certainly separated in the early Mesozoic. But as Moynihan has shown they still *share* four major categories of display! These are:

1. A display called the *dymantic,* in which two black spots appear on the body.
2. The *zebra stripes,* in which relatively close-packed transverse stripes appear on the body.
3. The *upward vee curls* in which two arms are raised and held back over the 'head', almost bull's horns fashion.
4. *Longitudinal streaks,* stripes disposed fore to aft.

Figure 7.6 Two octopus displays (from Moynihan, 1985). Eyes, eyespots and postures. Bottom left, the Dymantic display of *Octopus vulgaris*, and above, the display of *Octopus hummalincki*.

Moynihan argues that these displays must be ancestral because convergence on this scale is highly improbable. I find this a convincing argument. The 190 million year itch has not affected some really basic behaviour patterns. This raises a why? question. His answers are worth reading.

Moynihan has also (1985) used comparisons with the Cephalopoda to raise some fundamental questions in the field of animal communication. Until animals as capable of 'instant' and complex visual communication were studied in the field some of these questions were in a sense almost unthinkable. Cephalopods, particularly the social, free-swimming, non-cryptozoic squids and cuttlefishes, are equipped with the organic equivalent of a video display tube. Patterns of great complexity (Figure 7.6) can flash on their skin, flicker and change in instants. Nothing in the animal kingdom matches this. Moynihan and his co-worker Rodaniche (1977; 1982) have watched hours of 'continuous' pattern changes in the reef squid *Sepioteuthis sepiodea*. This mass of observations has led Moynihan (1985) to examine this communication in terms of rules and syntax. This treatment is challenging and provocative, one might even say 'far-out', but it is a true emergence of theory from a new line of comparative studies. It is the animals themselves that are 'far-out' and they may be the basis for opening a new door.

One of the things emerging from the cephalopod studies is that these animals exceed the quantitative limits of display repertoires known from vertebrate studies. The idea of limits on display repertoires was developed by Moynihan (1970) in a paper based on a comparative review. The theory and implications are interesting, but before dealing with them it is worth returning to the relatively huge repertoires of some cephalopods. Why do some of them exceed the 'norm' for vertebrates by as much as half an order of magnitude? Moynihan argues (1985) that some of the complex *ritualized* patterns classified as displays may, in fact, be what he has called anti-displays (1975). Anti-displays is a difficult term, but has been used to refer to complex adaptations that promote crypsis of, more generally, displays 'that are designed to impede the transfer of information to dangerous perceivers'. This is a most interesting extension of antipredator theory. I have suggested (Robinson, 1969b; 1981; 1985) that eucrypsis (camouflage *sensu stricto*) depends for its success in the *suppression of signals*, whereas all other systems depend on signalling either false information (plant-part and Batesian mimicry) or true information (aposematism). This scheme is repeated by Edmunds (1974) in his definitive 'Defence in animals'. Moynihan's cephalopod studies suggest a new extension of defence theory. An animal may produce utter confusion in a predator by very rapid changes in colours, patterns, shapes and textures. In a sense such a succession of anti-displays would constitute image suppression.

A further insight derived from comparative studies concerns the control, suppression, decay, disappearance and replacement of displays. After reviewing published accounts of the display repertoires of 62 species of vertebrate, mammals, birds and fishes, Moynihan (1970) concluded that, despite great differences in habits and social structure, the number of displays was remarkably similar. He points out that 'the usual range of displays would seem to be from approximately 15 to approximately 35'. This conclusion leads to a *why* question. The interesting answering suggestion is that repertoire ranges may be a diversity-dependent matter. The more displays the more some will inevitably be rare and unexpected. He also argues that the more diversity in the repertoire, the more diverse the form of each new display will have to be in order to be distinguishable, and therefore the more peculiar and disturbing it might be. This approach is full of possibilities for further study and analysis. It provides predictions about how and why displays should decline in a species history, and about why and with what they should be replaced. It is possibly a reflection on our ability to respond to rare, 'peculiar', and diverse ideas that its insights have been less used by others of the species ethologist than they should have been.

Another delightful example of the revelatory value of comparative studies relates to horned beetles. There has long been a mystery about the function of horns in beetles. Darwin was intrigued by the function of the horns of stag beetles that he found in Chile during the Beagle voyage. Opinions have differed ever since with hypotheses about horns as sex-attracting adornments and even functionless results of allometry. The problems involved in sorting out function were:

1. Lack of observations on living beetles.
2. The vast array of structures involved in 'horn' production (extensions of the head, mandibles, front legs, prothorax and so on).
3. The form of the 'horns' themselves, which is extraordinarily variable from the simple to the grotesque.
4. The fact that they occur in many unrelated beetle families.

Characteristically, the matter has been thoroughly dealt with by Eberhard (1979; 1980b, 1981) on the basis of studies of 17 species of horned beetles. Once the first mystery is elucidated the insight needed for the solution to others is accumulated incrementally. Here, in this study, comparison facilitates. Quite simply 'it has been established that far from being useless, beetle horns serve as effective weapons in contests between members of the same species over critical resources' (Eberhard 1980b). The horns are used in a remarkable variety of ways, some of which are shown in Figure 7.7. This is a clever study showing the power of ingenious, comparison-minded, investigation.

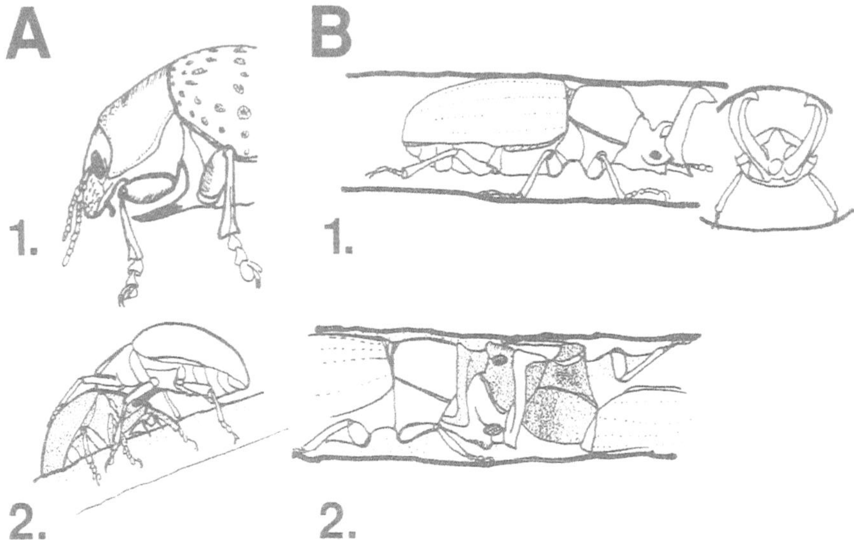

Figure 7.7 Horned beetles, after Eberhard. A1 shows the prothorax ventral horn of the beetle *Doryphora punctissima*, at first sight not a useable weapon because of its location and the fact that it does not extend beyond the head. A2 shows the use of this horn in combat. The unshaded individual is prying the other off the substrate. B1 shows the elongated antler-like horns of a darkling beetle close to the species *Molion muelleri*; since the beetle lives in tunnels it is difficult to guess at the utility of such jaws. B2 shows their use in combat. The more aggressive combatant turns through 180° and clamps its opponent behind the head.

CONCLUSION: PROBLEMS AND PROSPECTS

Problems

What problems exist in comparative studies? Some are common to all the evolutionary assumptions and interpretations derived from comparisons. Commonly, problems arise from comparing the wrong characters. This is true of the comparison between morphological characters that is the basis of most phylogeny construction. Such characters differ considerably in their variance from organ to organ and species to species. In the 'old days' expert taxonomists were expert because, after steeping themselves in their speciality for years, they had an inductive basis for assessing which characters were 'good'. Nowadays they confirm their intuitions with statistics, or perhaps even test their judgement that way. It is important that the characters chosen for comparison should be reliably species-specific. Since ethology's adolescent days of the 1960s our faith in the fixity of behaviour patterns has been somewhat eroded. For instance, Martin and May (1982) claim that 'Animal behaviour, both within and between species, tends to exhibit much greater variability than does morphology or physiology. This makes it difficult to reconstruct phylogenies from behaviour patterns.' This echoes some widespread scepticism. Recent emphases on studying intraspecific difference in behaviour, and the occurrence within a species of different evolutionary stable strategies, have tended to reduce the emphasis on broad comparative studies. This is in part a product of the fashions that sweep through ethology as well as other branches of science. It is also probably a reflection of changes in academic procedures and the state of the job market (these are not necessarily independent variables). Clutton-Brock and Harvey's (1984) review of comparative approaches to investigating adaptation provides a detailed critique of methods and comments: 'Our intention in reviewing the pitfalls of the comparative method is not to discourage the use of comparisons: many important questions can only be approached in this way. Moreover, it is worth remembering that while ideal comparisons are uncommon, ideal experiments which control for all possible confounding variables are rare.'

In conclusion, I am sure that we cannot afford to neglect the comparative method. The examples cited above, alone, and selective as they are, show what insights it can provide. I think they also show that in the hands of good naturalists, choices can be made about the bases of comparison that circumvent many of the problems that can otherwise bedevil the process. We can now, *a priori*, define some new caveats about the behavioural characters we use. For instance, the great revival of interest in sexual selection has been heuristic in many fields. We should now surely be convinced that where there is female choice there must be differences to be chosen between. This means that when we study courtship behaviour we must expect variability in those elements that can reflect, and be used by the female to detect, differences between males (and ultimately differences in their fitness). This means that, in phylogenic comparisons of courtship behaviour, we should be studying differences in strategies, not tactics. In particular, we should also compare the differences in strategies that make differences in tactics possible. (Eberhard (1985) has a marvellous book about differences in the form of male sex organs and the extent to which this depends on female preference. *Vive la différence* indeed!) It also seems to me that in general, behaviours associated with primary defence (preventing detection as prey, Robinson, 1969a) should be more reliably specific than those involved in secondary defence (i.e. those that operate to thwart attack after detection). The construction of prey-capture artefacts, and the types of predatory behaviour elicited by specific types of prey should also be strongly invariable. I have not even mentioned another field for comparative studies that seems to be of great potential, i.e. that of mapping the evolution of learning, intelligence, memory and consciousness. Griffin (1978, 1981, 1984) opened the doors to this for us. The prospects are exciting. We are, I hope, still curious naturalists all.

The use of comparative methods in studies of evolution, and particularly, for me at least, in studies of the Paley phenomenon, is a great game (in the sense of Kipling's *Kim*), fascinating, tantalizing, exciting and ultimately satisfying. Niko Tinbergen delighted in the wonders produced by the evolutionary process and the great game of discovery. I hope we continue to play at it.

REFERENCES

Baerends, G.P. and Barends-van-Roon, J.M. (1950) An introduction to the study of the ethology of cichlid fishes. *Behaviour*, Suppl **1**, 1–242.

Blest, A.D. (1957a) The function of eye-spot patterns in the Lepidoptera. *Behaviour*, **11**, 209–56.

Blest, A.D. (1957b) The evolution of protective displays in the *Saturnioidea* and *Sphingidae* (Lepidoptera). *Behaviour*, **11**, 257–309.

Blest, A.D. (1963a) Relations between moths and predators. *Nature*, Lond., **197**, 1046–7.

Blest, A.D. (1963b) Longevity, palatability and sound production in some New World arctiid and ctenuchid moths. *Zoologica*, **49**, 161–81.

Cloarec, A. (1988) Behavioral adaptations to aquatic life in insects, in *Advances in the Study of Behavior*, (eds. J.S. Rosenblatt, C. Beer, M-C. Busnel, P.J.B. Slater), Academic Press, New York, pp. 99–151.

Coddington, J. (1986) The monophyletic origin of the orb web, in *Spiders, webs, behavior and evolution*, (ed W.A. Shear) Stanford University Press, Stanford, California, pp. 319–63.

Clutton-Brock, T.H. and Harvey, P.H. (1984) Comparative approaches to investigating adaptation, in *Behavioural Ecology*, (eds J.R. Krebs and N.B. Davies), Sinauer Associates Inc., Cambridge, pp. 7–29.

Clyne, D. (1973) Notes on the web of *Poecilopachys australasia* (Griffin and Pidgeon 1833) (Araneidae: Argiopidae). *Australia Entomologist's Magazine*, **1**, 23–29.

Crane, J. (1952) A comparative study of the innate defensive behaviour of Trinidad mantids (Orthoptera, Mantoidea). *Zoologica*, **37**, 259–93.

Dawkins, R. (1986) *The Blind Watchmaker*, W.W. Norton, New York and London.

De Ruiter, L. (1952) Some experiments on the camouflage of stick caterpillars. *Behaviour*, **4**, 222–32.

De Ruiter, L. (1955) Countershading in caterpillars. *Archs. néerlandaise de Zoologie*, **11**, 1–17.
Eberhard, W.G. (1967) Attack behavior of diguetid spiders and the origin of prey-wrapping in spiders. *Psyche*, **74**, 173–81.
Eberhard, W.G. (1975) The 'inverted ladder' orb web of *Scoloderus sp*, and the intermediate orb of *Eustala sp.* (Araneae: Araneidae). *Journal of Natural History*, **9**, 93–106.
Eberhard, W.G. (1977) Aggressive chemical mimicry by a bolas spider. *Science*, **198**, 1173–75.
Eberhard, W.G. (1979) The function of horns in *Podischnus agenor* and other beetles, in *Sexual selection and reproductive competition in insects*, (eds M. Blum and N. Blum), Academic Press, New York, pp. 231–58.
Eberhard, W.G. (1980a) The natural history and behavior of the bolas spider, *Mastophora dizzydeani* sp. N. (Araneidae). *Psyche*, **87**, 143–69.
Eberhard, W.G. (1980b) Horned beetles. *Scientific American*, **242**, 166–82.
Eberhard, W.G. (1981) The natural history of *Doryphora* sp. (Coleoptera, Chrysomelidae) and the function of its sternal horn. *Annals of Entomological Society of America*, **74**, 445–48.
Eberhard, W.G. (1982) Behavioral characteristics for the higher classification of orb-weaving spiders. *Evolution*, **36**, 1067–95.
Eberhard, W.G. (1985) *Sexual selection and animal genitalia*, Harvard University Press, Cambridge, Mass.
Eberhard, W.G. (1986) Effects of orb-web geometry on prey interception and retention, in *Spiders, webs, behavior and evolution*, (ed W.A. Shear) Stanford University Press, Stanford, Calif. pp. 70–100.
Edmunds, M. (1972) Defensive behaviour in Ghanaian praying mantids. *Zoological Journal of the Linnean Society*, **51**, 1–32.
Edmunds, M. (1974) *Defence in Animals*, Longman, London.
Edmunds, M. (1976) The defensive behaviour of Ghanaian praying mantids with a discussion of territoriality. *Zoological Journal of the Linnean Society*, **58**, 1–37.
Eisner, T., Alsop, R., Ettershank, G. (1964) Adhesiveness of spider silk. *Science*, **146**, 1058–61.
Eldredge, N. (1985a) *Time frames: The rethinking of Darwinian evolution and the theory of punctuated equilibria*. Simon and Schuster, New York.
Eldredge, N. (1985b) *Unfinished synthesis: Biological hierarchies and modern evolutionary thought*. Oxford University Press, New York.
Eldredge, N. (1989) *Macro-evolutionary Dynamics*. McGraw-Hill, New York.
Elton, S.S. (1973) The structure of invertebrate populations inside neotropical rainforest. *Journal Animal Ecology*, **42**, 55–104.
Erwin, T. (1982) Tropical forests: their richness in coleoptera and other arthropod species. *The Coleopterist's Bulletin*, **34**, 305–22.
Griffin, D.R. (1978) Prospects for a cognitive ethology. *Behavioral and Brain Sciences*, **1**, 527–38.
Griffin, D.R. (1981) *The Question of Animal Awareness*. Rockefeller University Press, New York.
Griffin, D.R. (1984) *Animal Thinking*. Harvard University Press, Cambridge, MA, USA; London, England
Gould, S.J. (1980) Is a new and general theory of evolution emerging? *Paleobiology*, **6**, 119–30.

Gould, S.J. and Eldredge, N. (1977) Punctuated equilibria: The tempo and mode of evolution reconsidered. *Paleobiology*, **3**, 115–51.

Kranz, K.(1982) A note on the tail hairs from a pygmy hippopotamus (*Choeropsis liberiensis*). *Zoo Biology*, **1**, 237–41.

Hinde, R.A. and Tinbergen, N. (1958) The comparative study of species-specific behaviour, in *Behaviour and Evolution*, (eds A. Roe and G.G. Simpson, Yale University Press, Hartford, pp. 251–68.

Mayr, E.(1982) *The Growth of Biological Thought*. Harvard University Press, Cambridge, MA, USA, London, England.

Mayr, E. (1988) *Towards a new philosophy of biology*, Harvard University Press, Cambridge, Mass.

Martin, R.D. and May, R. (1982) Outward signs of breeding, in *Evolution now*, (ed J.M. Smith) Freeman, San Francisco.

Moynihan, M.H. (1959) *A revision of the family Laridae* (Aves), *American Museum Novitates*, **1928**, 1–42.

Moynihan, M.H. (1962) Hostile and sexual behavior patterns of South American and Pacific Laridae. *Behaviour, Suppl*, **8**, 1–365.

Moynihan, M.H. (1970) The control, suppression, decay, disappearance and replacement of displays. *Journal of Theoretical Biology*, **29**, 85–112.

Moynihan, M.H. (1975) Conservatism of displays and comparable stereotyped patterns among cephalopods, in *Function and evolution in behaviour*, (eds G. Baerends, C. Beer, A. Manning), Oxford University Press, Oxford, pp. 276–91.

Moynihan, M.H. (1983a) Notes on the behavior of *Europrymna scolopes* (Cephalopoda, Sepiolidae). *Behaviour*, **85**, 25–41.

Moynihan, M.H. (1983b) Notes on the behavior of *Idiosepius pygmaeus* (Cephalopoda, Idiosepiidae). *Behaviour*, **85**, 42–57.

Moynihan, M.H. (1985) *Communication and noncommunication in cephalopods*, Indiana University Press, Bloomington, Indiana.

Moynihan, M.H. and Rodaniche, A.F. (1977) Communication, crypsis and mimicry among cephalopods, in *How animals communicate*, (ed T. Sebeok) Indiana University Press, Bloomington, Indiana, pp. 293–302.

Moynihan, M.H. and Rodaniche, A.F. (1982) The behavior and natural history of the Caribbean reef squid *Sepioteuthis sepioidea*. *Zeitschrift für Tierpsychologie Advances in Ethology*, **25**, 1–150.

Ridley, M. (1983) *The explanation of organic diversity*, Oxford University Press, Oxford.

Robinson, M.H. (1964) The Javanese stick insect, *Orxines macklotti* De Haan (Phasmatodea, Phasmidae). *The Entomologist's Monthly Magazine*, **100**, 254–59.

Robinson, M.H. (1968a) The defensive behavior of the Javanese stick insect *Orxines macklotti* De Haan, with a note on the startle display of *Metriotes diocles* Westw. (Phasmatodea, Phasmidae). *The Entomologist's Monthly Magazine*. **104**, 46–51.

Robinson, M.H. (1968b) The startle display of *Balboa tibialis* (Brunner), (Orth., Tettigoniidae). *The Entomologist's Monthly Magazine*, **104**, 88–90.

Robinson, M.H. (1968c) The defensive behavior of *Pterinoxylus spinulosus* Redtenbacher, a winged stick insect from Panama (Phasmatodea). *Psyche*, **75**, 195–207.

Robinson, M.H. (1968d) The defensive behavior of the stick insect *Oncotophasma martini* (Griffini). (Orthopotera: Phasmatidae). *Proceedings of the Royal Entomological Society, London*, **43**, 183–87.

Robinson, M.H. (1969a) Defenses against visually hunting predators. *Evolutionary Biology*, **3**, 225–59.
Robinson, M.H. (1969b) The defensive behavior of some orthopteroid insects from Panama. *Transactions of the Royal Entomological Society, London*, **121**, 281–303.
Robinson, M.H. (1969d) The predatory behavior of *Argiope argentata* (Fabricius). *American Zoologist*, **9**, 161–73.
Robinson, M.H. (1970a) Animals that mimic parts of plants. *Morris Arboretum Bulletin*, **21**, 51–58.
Robinson, M.H. (1970b) Insect antipredator adaptations and the behavior of predatory primates. *Act. IV Congr. Latin Zool*, **2**, 811–36.
Robinson, M.H. (1973) The evolution of cryptic postures in insects, with special reference to some New Guinea tettigoniids (Orthoptera). *Psyche*, **80**, 159–65.
Robinson, M.H. (1975) The evolution of predatory behaviour in araneid spiders, in *Essays on the evolution and function of behaviour*, (eds G. Baerends, C. Beer, and A. Manning), Oxford University Press, Oxford, pp. 292–312.
Robinson, M.H. (1977) Informational complexity in tropical rain forest habitats and the origins of intelligence. *Actas del IV Symposium Internacional de Ecologia Tropical Panama*, **1**, 148–68.
Robinson, M.H. (1978) Tropical biology; is it real? *Tropical Ecology*, **19**, 30–50.
Robinson, M.H. (1979) By dawn's early light: matutinal mating and sex attractants in a neotropical praying mantis. *Science*, **205**, 825–27.
Robinson, M.H. (1981) A stick is a stick is a stick: on the definition of mimicry. *Biological Journal of the Linnean Society of London*, **16**, 1–6.
Robinson, M.H. (1982) Courtship and mating behaviour in spiders. *Annual Review of Entomology*, 1–20.
Robinson, M.H. (1985) Predator–prey interactions, informational complexity, and the origins of intelligence. *Journal of the Washington Academy of Sciences*, **75**, (**4**), 91–104.
Robinson, M.H., Mirik, H. and Turner, O. (1969c) The predatory behavior of some araneid spiders and the origin of immobilization wrapping. *Psyche*, **76**, 487–501.
Robinson, M.H. and T. Pratt (1976) The phenology of *Hexacentrus mundus* at Wau, Papua New Guinea (Orthoptera: Tettigoniidae). *Psyche*, **82**, 315–23.
Robinson, M.H. and Robinson, B. (1972) The structure, possible function and origin of the remarkable ladder-web produced by a New Guinea orb-web spider. *Journal of Natural History*, **6**,687–94.
Robinson, M.H. and Robinson, B. (1975) Evolution beyond the orb-web, the web of *Pasilobus* sp.: its structure, construction and function. *Zoological Journal of the Linnean Society of London*, **576**, 301–14.
Robinson, M.H. and Robinson, B. (1978) The evolution of courtship systems in tropical araneid spiders *Symposium of the Zoological Society of London*, **42**, 17–29.
Robinson, M.H. and Robinson, B. (1980) Comparative studies of the courtship and mating behavior of tropical araneid spiders. *Pacific Insects Monographs*, **36**.
Shear, W.A. (ed) (1986) *Spiders, webs, behavior and evolution*, Stanford University Press, Stanford, CA.

Stowe, M.K. (1986) Prey specialization in the Araneidae, in *Spiders, webs, behavior and evolution*, (ed W.A. Shear), Stanford University Press, Stanford, Calif., pp. 101–132.

Tinbergen, N. (1951) *The study of instinct*, Oxford University Press, Oxford.

Tinbergen, N. (1954) The origin and evolution of courtship and threat display, in *Evolution as a process*, (eds. A.C. Hardy, J.S. Huxley and E.B. Ford) Allen and Unwin, London, pp. 233–50.

Tinbergen, N. (1959a) Behavior, systematics and natural selection. *Ibis*, **101**, 418–30.

Tinbergen, N. (1959b) Comparative studies of the behaviour of gulls (Laridae): a progress report. *Behaviour*, **15**, 1–70.

Tinbergen, N. (1960) The evolution of behavior in gulls. *Scientific American*, **203**, 118–30.

Tinbergen, N. (1962) The evolution of animal communication, *Symposium of the Zoological Society, London*, **8**, 1–8.

Tinbergen, N. (1963) On aims and methods in ethology. *Zeitschrift für Tierpsychologie*, **20**, 410–33.

Tinbergen, N. (1964) On adaptive radiation in gulls (Tribe Larini). *Zool. Meded. Leiden*, **34**, 209–23.

Tinbergen, N. (1965a) Behavior and natural selection, in *Ideas in modern biology, Proceedings XVI International Congress Zool.*, (ed J.A. Moore) Natural History Press, New York, pp. 521–42.

Tinbergen, N. (1965b) Some recent studies of the evolution of sexual behavior, in *Sex and behavior*, (ed F.A. Beach), Wiley, New York, pp. 1–34.

Tinbergen, N. (1972) *The animal in its world*, Vol. 1. Allen and Unwin, London.

Tinbergen, N., Broekhuysen, G.J., Feekes, F. *et al.* (1962) Egg-shell removal by the black-headed gull, *Larus ridibundus* L., a behaviour component of camouflage. *Behaviour*, **9**, 74–117.

8

The Tinbergen legacy in photography and film

LARY SHAFFER

Niko Tinbergen always felt that something was wrong because he was able to spend much of his professional life doing things that he enjoyed enormously: watching and photographing animals, and trying to work out the interaction between behaviour and evolution. He wrote an elegant justification of his activities near the end of 'Curious naturalists' (Tinbergen, 1958):

> 'It seems to me that no man need be ashamed of being curious about nature. It could even be argued that this is what he got his brains for and that no greater insult to nature and to oneself is possible than to be indifferent to nature. There are occupations of decidedly lesser standing.'

In return for the excitement and fun he was having, Niko felt that he had an obligation to the public who, ultimately, paid for his work. Throughout his professional life, he placed a very high value on communicating the results of his studies to the general public. He said 'I try to impress on my students that half their work is communication. Science is a social effort, and scientists must adjust to the public. If people don't want to read your work, your whole effort, and all the money that went into it has been lost.' (Hall, 1974).

He felt it was an important responsibility to present research in forms which would attract and engage people. Having disdain for the stilted style of the scientific journals, he always encouraged his students to write in the first person and to minimize jargon. An important aspect of the Tinbergen legacy in animal behaviour is the example he set in using film and photography to describe his findings.

Niko was an artist as well as a scientist. Although he rarely indulged himself, when he took the time, he showed considerable talent at pencil sketching. This skill was a useful technical tool for him as he worked in the field because he would often make simple sketches of behaviours to illustrate his notes. These quick drawings displayed the essence of the behaviour with a remarkable economy of line. Niko's drawings scattered throughout 'Curious naturalists' (Tinbergen, 1958) and 'The Herring Gull's World' (Tinbergen, 1953) are splendid examples of this. Niko described how, as a boy, he yearned to produce photographs which could accurately capture much more detail than the drawings he was making.

Niko had a lifelong interest in photography and his artistic gifts are evident in the wonderful photographs which illustrate his books. His lifetime spanned a time of rapid change and improvement in photography and Niko did his best - with very limited funds at first - to take advantage of new developments as they came along. For much of his life his photographic skill seems to have been hampered by the equipment which was available to him. He was ready to fly before his equipment knew how to walk.

His earliest photographs were taken on clumsy 6½ × 8½ inch glass plates, called 'whole plates'. He often amusingly talked about the struggles involved in trying to photograph animals with the primitive plate cameras which accompanied the fragile glass negatives. The number of negatives which could be exposed on a given day was limited by the number of plate holders Niko owned. By careful saving he finally managed to acquire three double holders which took a plate in each side, permitting six exposures. Once he had exposed these plates, it was necessary to return home to develop them and reload the plate holder. It always seemed to Niko that the most exciting photographs were on the plates which would get broken on the bicycle ride home.

There were other problems related to trying to photograph natural history, and particularly herring gulls, with a plate camera. Perhaps first among these problems was trying to use such a bulky camera in a hide. Because the viewfinder was on top of the camera and it was necessary for both viewfinder and lens to have a view from the hide, it was almost impossible to disguise the camera without alarming the gulls. Most of Niko's plate camera pictures from hides were taken by simply waiting until something happened in front of the camera then snapping the shutter. The greatest drawback of the plate camera was that it was not reflex - the photographer did not look through the lens but rather through a separate viewfinder. Because this viewfinder only approximated the scene that would be photographed, focus was a matter of guesswork. Even for those pictures which happened to be properly focused important parts of the action would sometimes occur beyond the edge of the negative. The plate camera could be focused through the lens by removing the plate holder, but by the time the plate holder was reinserted and its cover removed, the reason for the photograph would often be gone.

Niko developed an additional strategy for photographing the gulls with the plate camera, which involved camouflaging the camera under an old basket or something similar near to the probable site of action. The remote camera was watched from a hide. At just the right moment, when the right behaviour was happening, in what seemed to be the right distance in front of the camera, Niko would release the shutter with a long bulb-operated release. This yielded larger pictures of the gulls, allowing better prints to be made, but it had the drawback that only one picture could be taken at a time. Following the exposure, Niko would have to get out of the hide, disturb the animals, and go to the camera to insert another plate. Niko used his detailed knowledge of the gulls in order to successfully set up the camera just where the animal was likely to be at the moment that something interesting was about to happen. A problem for him however, was that the gulls liked to use the camera basket as a lookout post and much of the interesting action took place on top of the camera.

Niko showed great determination in those days and although most of the old glass plate negatives did not result in anything useful, a few are remarkable illustrations of behaviour.

Before the war, large boxy Soho reflex-type cameras became available and Niko used one of these for a short time. This camera used film which was 3¼ by 4¼ inches. This machine had at least three advantages over the plate camera. It could use unbreakable celluloid cut film and it came with a 'bookform' film holder which allowed more exposures to be carried to the field and more rapid reloading for the taking of the next exposure. More important, as a single lens reflex camera it allowed Niko to focus and compose the scene through the lens that was taking the picture. This solved some of the problems of use in a hide because all but the lens of the camera could be obscured by camouflage and slow, careful movements of the camera became possible. The drawback was that before a picture could be taken, the larger mirror which directed the image to the viewfinder had to be winched up manually by pulling a string to move it out of the path between the lens and the film. It was necessary for Niko to anticipate the pattern of behaviour so that the mirror winching could be initiated before the desired behaviour happened. Niko often remarked that he learned a great deal about the patterns of behaviour by trying to photograph them with this primitive camera. At the end of the exposure, the reflex mirror would drop to the viewfinding position on its own with a clang that would upset the animals for several minutes.

For the expedition that Niko and his wife Lies took to Greenland in 1932, Niko used a Rolleiflex. This was a twin lens reflex camera having two matched lenses: one for viewfinding and one for photographing. The Rolleiflex was much more portable than the earlier reflexes and had the added advantage that it used roll film which could easily be reloaded in the field. The relative ease of using this camera, as well as its sharper lenses and higher resolution film enabled Niko to return with hauntingly beautiful photographs of Greenland and its inhabitants, some of which are reproduced in 'Curious naturalists' (Tinbergen, 1958). The tameness of the snow buntings and red-necked phalaropes in virtually uninhabited parts of Greenland allowed Niko to take good close-up pictures without a hide, using the short lenses of the Rolleiflex.

After the war Niko brought an Alpa reflex which was one of the early small 35 mm single lens reflexes. The lens on the Alpa was interchangeable but Niko did not have sufficient funds to buy the expensive telephoto lenses which were available. However, at this time, there were a number of war department surplus telephoto lenses available which had been part of aerial photography cameras. Niko obtained two of these magnificently sharp lenses, a 280 mm and a 340 mm and had them adapted to the Alpa in the Zoology Department workshop in Leiden. Robust, variable length barrels were fabricated from heavy 3-inch brass pipe and fitted to the lenses to permit focusing. The laboratory technicians used a 1½ inch pipe thread (the only large diameter threading equipment in the workshop) to attach the enormously heavy lenses to a brass clamp which wrapped around the Alpa body. These mammoth lenses were attached to the tripod with the tiny Alpa hanging on at the back.

The adaptation was crude but the results were not. Many of the memorable pictures of gulls in 'The Herring Gull's World' (Tinbergen, 1953) and many of the wonderful animal pictures, such as those of the kittiwakes in 'Curious naturalists' (Tinbergen, 1958) were taken with these brass cannons. After he moved to Britain, Niko eventually retired the Alpa and purchased one of the early Nikon F single lens reflexes. With this camera, several Nikkor lenses and a 400 mm Novoflex, Niko took many of the pictures for 'Tracks' (Ennion and Tinbergen, 1967) and for 'Signals for survival (Tinbergen *et al.*, 1970). At last Niko had the equipment required to reach out into the world of gulls and other animals and the result is that the 'Signals' book tells the story of the gulls with few words but with many excellent pictures.

Perhaps because of his early training with glass plates, Niko always shot film economically. He would return from a hide session with three rolls of exposed film saying that he had 'just blasted away at everything.' In the darkroom, he had a good eye for finding the right negative and made excellent prints with casual abandon. After making only one small exposure test strip under the enlarger, he only rarely made the prints which needed to be rejected because of total quality. In the darkroom he would grab prints from the developing trays with his hands, eschewing print tongs as being too cumbersome. I always found it alarming to watch him leaning over the developing trays with a long and delicate ash hanging from the hand-rolled cigarette clenched in his lips. While I never actually saw one of these ashes fall into the developer, or, in the field, into an open camera back, I always expected to.

Niko was a pioneer in the making of natural history films which tell a research story. While biologists have been making natural history films about their own research since motion pictures first appeared, until recently very few of these films were sufficiently polished to bring the research stories to a wide public. A noteworthy early exception was Julian Huxley's 1937 Academy Award winning gannet film.

Niko started taking motion pictures while still in Holland and produced a film about the courtship of the three-spined stickleback, as well as a film giving a composite view of research and field camp life in Hulshorst. Niko also tried to film herrring gulls, but judging from the surviving scraps of film, camera limitations made it uneconomic to attempt a gull film at this point. As far as I can ascertain, these earliest films were photographed with a Kodak 16 mm, non-reflex, spring-driven camera with a single short lens. The viewfinder was on the top of this camera and so difficulties with the viewfinding showed themselves in the vertical dimension, with the tendecy of heads or feet to be cut off. The difficulty in getting moving, non-human animals in focus was very great and much of the stickleback film avoids close-up shots, in which this difficulty was magnified.

The black and white, silent stickleback and Hulshorst films were given a professional polish by the addition of title cards to help tell the story. They were used extensively as teaching films in the Zoology department at Leiden. When Niko moved to Oxford, the stickleback film was handed over to the British Film Institute for educational distribution to schools. By the time they were retired, the BFI prints of the stickleback film were worn ragged, attesting to the numbers of showings they must have had.

Upon Niko's arrival in Oxford, Alister Hardy helped to make possible the purchase of a Bell & Howell HR-1 which was a sturdy spring-wound 16 mm camera which had the advantage of having a three-lens turret. In these days before zoom lenses, this turret permitted three different focal length lenses to be used in succession by rotating the turret until another lens came into the breech. The range of the turret-mounted lenses was limited because a telephoto lens mounted on the turret would be part of the scene being photographed by wider angle lenses. Although the Bell & Howell was not a reflex camera, its makers had provided a small focusing peephole near the edge of the lens turret. A lens was rotated until it was adjacent to this peephole, through which it could be focused on the subject. Then it was rotated back to the talking position before filming could begin. The viewfinder had an adequate adjustment for parallax.

Niko was determined to make a series of films for use in teaching at Oxford. He realized the potential which film offered to bring the excitement of methods and results of the field camp work back to the unfortunates who had been left behind in the lab. He started to work on this project shortly after arrival in Oxford. Surviving scraps of film show the attempts to study gulls on Scolt Head Island, where the spring tide was so high that it covered much of the breeding colony, leaving gulls swimming above their nests, waiting for the tide to recede.

The earliest of Niko's Oxford teaching films stressed the *results* of the research, rather than the *process* of research. Among the first were a film about Esther Cullen's kittiwakes on the Farne Islands, and a film showing Bernard Kettlewell's classic experiments on industrial melanism in the Peppered Moth. According to Niko, the Peppered Moth film was made as a result of doubts, in some quarters, that birds could possibly be responsible for the differential predation of black or pale-chequered moths. Working together, Niko and Kettlewell were able to produce convincing film evidence with good close-up shots of birds grabbing poorly camouflaged moths from tree trunks (Tinbergen, 1958). The kittiwake and the moths were ideal subjects for filming with the Bell & Howell, because the action occurred in quite restricted areas (small ledges or small sections of tree trunk) which could be focused upon in advance using the little peephole. Niko had acquired a 150 mm telephoto lens with which he was able, for the first time, to get good motion picture close-ups of wild animals (Figure 8.1).

The Bell & Howell went with Niko to the Ravenglass field camp where he made a series of teaching films. The earliest of these was a descriptive film about the courtship of the black-headed gull but his subsequent efforts were what he called 'research-in-action' films. These showed the researchers and details of their experimental manipulations as well as the responses of the animals involved. This tactic had been

Figure 8.1.

presaged in the old Hulshorst film, but at Ravenglass it became the major focus of Niko's film-making efforts.

At the time that Niko was making these research-in-action teaching films, he met Hugh Falkus. Hugh was working for Border Television making a little series called 'Five Minutes with Hugh Falkus' in which he drifted around the border country finding stories of local interest which could be told in five minutes. While travelling around, Hugh had heard about the work of the Oxford Behaviour group at Ravenglass. He visited, talked and ultimately made four of the little programmes about research there.

Niko was fascinated by the extent to which television could bring animal behaviour research to a very large audience, and he agreed to work with the BBC to make three research-in-action programs for the 'Look' series. These half-hour, black and white programmes ('The sign readers', 'The beachcombers' and 'The gull watchers') contain quite a bit of Niko's animal behaviour film as well as some interview material filmed by the BBC.

At the same time, Niko was turning out memorable silent teaching films in colour on eggshell removal in the black-headed gull, search image in the carrion crow, and the breeding biology of the oystercatcher

Figure 8.2 Niko Tinbergen filming oystercatcher chicks in 1971.

(Figure 8.2). Niko also travelled to the Bass Rock to film the gannet studies of Bryan Nelson. These films brought field work alive for a generation of students.

In 1966 Niko began another collaboration with Hugh Falkus to film the life story of the lesser black-backed gull on Walney Island. This film, 'Signals for survival', was commissioned by the BBC for the 'World about us' series. By this time Niko had acquired a 16 mm, reflex Bolex motion picture camera. Armed with the telephoto lenses he had used with his Nikon, as well as a little Pan Cinor zoom lens he had the capability to catch the fast-moving action which punctuates this film. With the 'Signals' film the equipment ceased to limit his talent and the resulting programme is a film of unusual beauty as well as lasting educational value (Figure 8.3).

Niko learned about the syntax of films while filming 'Signals' with Hugh Falkus. Hugh taught Niko about the use of long shots, close-ups and establishing shots to tell the story visually. Niko's previous films had told their story clearly, but the story was told with film clips which were, effectively, animated slides. Because of the limitations of the non-reflex cameras, much of the early film that Niko took consisted of multiple shots of the same animal against the same

Figure 8.3 Niko Tinbergen filming 'Behaviour and survival' in 1972.

background at the same magnification in the film frame. It is very difficult for a film editor to make a smooth visual story with this kind of material. When one segment ends and another begins the animal on the screen seems to jump suddenly because the background and magnification are the same but the animal has moved between one shot and another. This kind of film edit is called, with justification, jump cut. If the editor has a variety of different magnifications or different camera angles with which to work, then the film can be fitted together like a linear jigsaw puzzle. Two segments which would jump if hitched together directly can be used near to each other without the disruptive visual jump, if separated by a close-up.

Niko knew the gulls and he knew their story. Hugh knew how to build a film. Their collaboration had its rough edges, but it worked because each had respect for the expertise of the other. For the two seasons during which 'Signals' was being made, Niko trooped out to the hides with Hugh's lists of specific shots which would be needed: close up head and shoulders of gull facing left; medium shot of mated pair with grass background; and so on. Talking of this time, Niko often said that it was very difficult for him to concentrate on getting the specific shots which were needed when, at the same time, other exciting things would be going on within camera range. He gave in at times and some of these unscripted happenings greatly enriched the film.

'Signals' was first transmitted in 1968 and in 1969 it won the prestigious Italia Prize for television documentaries. On the heels of this international recognition, Niko was approached by the BBC and Time/Life who wanted to make a series of half-hour television programmes on animal behaviour to be called 'Behaviour and survival'. They wanted Niko to supervise the series and make as many of the programmes as he could. In the end, Niko did little filming for this series, although some of the films used old material dating back to the Ravenglass days. Niko wrote, edited and directed the filming of three of these programmes, 'The mussel specialist', 'Tracking', and an introductory programme. For the other ten programmes he was involved in the original treatment and plan and then came in again after filming to polish off the editing and the script. These programmes have worn well and nearly twenty years later they are still running in some television markets. They also have seen very wide educational distribution.

There were two themes to which Niko frequently returned when talking about natural history photography and film making. First, he had a deep commitment to knowing the animal well, so that his films and photographs, which must be selective, showed an accurate picture of animal behaviour. Second, although Niko had an extraordinary sense of pictorial beauty and composition, he disdained natural history films or books which were, in his words 'pretty-pretty', with no substance. In this category he placed films which were nothing more than montages of pictures of animals with no attempt to build a story. In all his work, the story was the important thing. His films and his books are scientific and artistic works, but they are also deftly told stories and it is for this reason that they are so memorable.

REFERENCES

Ennion, E. and Tinbergen, N. (1967) *Tracks*, Oxford University Press, Oxford.

Hall, E. (1974) A conversation with Nobel Prize winner Niko Tinbergen. *Psychology Today*, **7**, 65–80.

Tinbergen, N. (1953). *The herring gull's world*, Collins, London.

Tinbergen, N. (1958) *Curious naturalists*, Country Life, London.

Tinbergen, N., Falkus, H. and Ennion, E. (1970) *Signals for survival*, Oxford University Press, Oxford.

Afterword

AUBREY MANNING

This meeting had a strong family feeling about it. Most people attending knew Niko Tinbergen and there was a unique gathering of his old students, although we sadly missed Mike Cullen who was such a key figure in the group for many years. Our thanks go to Marian Dawkins, Tim Halliday and Richard Dawkins for arranging this celebratory reunion. It was particularly nice for us to be joined by members of the real Tinbergen family. It cannot always have been easy to have famous parents. The children must have been gull-orphans from time to time and they also had to put with a stream of visitors competing for their parents' attention. However, they could not fail to have recognized the esteem and deep affection that all our extended ethological family have always felt for Niko and Lies.

Most of the papers read here included some personal reminiscence along with science. We would all acknowledge that the Tinbergen legacy had a distinctive style about it, as well as its distinctive biological content. There was the marvellous atmosphere of 'Friday evenings' at the Tinbergens' house, for example, which provided a mixture of informality and rigour. Rigour meant that Niko obstinately refused to let discussion move on until he had fully understood the argument – a lesson I have tried hard to live up to when the temptation is just to let things go and hope you'll be able to work it out later.

The manifestation of his style which I shall most cherish derives from vivid memories of one or two walks in Wytham, when Niko came to visit my study area. He offered a spontaneous running commentary on the natural world around – the timing of nectar secretion from a flower, the flight intention movements of a bird, the nesting burrow of a solitary bee – everything caught his eye and he could fit into a pattern of life; I can call it no less. Once or twice I could show *him* something new with my bumble-bees and share his utter delight for a new insight into the way their world is organized. We can all understand why he called his essay of scientific autobiography* 'Watching and wondering' - this latter word to be taken in both senses.

As we know, he was modest in a most extreme way, and he remained genuinely amazed at the prestige he commanded in the world of science. Desmond Morris recounts how when the Nobel Prize was announced, his first response was astonished modesty but with an immediate afterthought - 'it'll be good for ethology'. Of course it was, but through Niko himself, for it was he who led the ethological approach into its fullest maturity. So many of his ideas have entered into our way of thinking as all the papers in this volume bear evidence.

We were saddened by the death of Lies just a few days before our meeting. All of us remember them together. Lies worked with him in the early days in Greenland, she was there at every Friday evening (often with the most telling interjections) and they collaborated for their later research on childhood autism. Whenever I visited them over these past few years we always talked about this work which they did together. Their commentary was as lively (and often acerbic!) as ever and this is how I shall remember them.

In that autobiographical essay, Niko wrote 'But what men in my position can truly say is that they have been privileged to have lived in such an interesting time and to have witnessed and assisted both in the birth, or rather the rebirth, and the coming-of-age of a fascinating new branch of biology.' Here is the characteristic modesty again, but people in our position recognize how privileged we have been to be members of that generation which has learnt from him directly. We are all beneficiaries of The Tinbergen Legacy.

* Tinbergen, N. (1985) Watching and wondering, in *Studying animal behavior: autobiographies of the founders* (ed D.A. Dewsbury) University of Chicago Press, Chicago and London, pp. 431–63.

Index

GPSR Compliance
The European Union's (EU) General Product Safety Regulation (GPSR) is a set of rules that requires consumer products to be safe and our obligations to ensure this.

If you have any concerns about our products, you can contact us on

ProductSafety@springernature.com

In case Publisher is established outside the EU, the EU authorized representative is:

Springer Nature Customer Service Center GmbH
Europaplatz 3
69115 Heidelberg, Germany

www.ingramcontent.com/pod-product-compliance
Ingram Content Group UK Ltd.
Pitfield, Milton Keynes, MK11 3LW, UK
UKHW021333070726
13610UKWH00011B/61

* 9 7 8 9 4 0 1 7 3 7 4 9 4 *